PHOTOSYNTHESIS

CELL BIOLOGY: A SERIES OF MONOGRAPHS

E. Edward Bittar, Series Editor

PHOTOSYNTHESIS

CHRISTINE H. FOYER
Research Institute for Photosynthesis
Sheffield University

A WILEY-INTERSCIENCE PUBLICATION
JOHN WILEY & SONS
New York • Chichester • Brisbane • Toronto • Singapore

10/1985
Biol.

Library of Congress Cataloging in Publication Data:

Foyer, Christine H.
  Photosynthesis.

  (Wiley series on cell biology ; v. 1)
  "A Wiley-Interscience publication."
  Includes bibliographies and index.
  1. Photosynthesis. I. Title. II. Series.

QK882.F64 1984 581.1′3342 83-21764
ISBN 0-471-86473-0

SERIES PREFACE

The aim of the Cell Biology Series is to focus attention upon basic problems and show that cell biology as a discipline is gradually maturing. In its largest aim, each monograph seeks to be readable and informative, scholarly, and the work of a single mind. In general, the topics chosen deal with major contemporary issues. Together they represent a rather large domain whose importance has grown enormously in the course of the last generation. The introduction of new techniques has no doubt ushered in a small revolution in cell biology. However, we still know very little about the cell as an ordered structure. As will become abundantly clear to the reader, real progress is not just a matter of progress of technique but also a matter of close interaction between advances in different fields of study, as well as genesis of new approaches and generalized concepts.

E. EDWARD BITTAR

Madison, Wisconsin
January 1984

PREFACE

There are but a few textbooks available on photosynthesis. These are often expert studies on specific aspects of the overall process centered on the major fields of interest of the authors. Alternatively, there are several collective works consisting of highly specialized individual reviews of various topics related to photosynthesis. These are invaluable to the research biochemist. Here it was my intention to provide a concise overall review of the photosynthetic pathways and bridge the gap between the weighty review compilations and the detailed insular texts. I have endeavored to discuss the many processes that are involved in photosynthesis, providing an accurate contemporary account of the pathways, their regulation, and interrelationships. Of necessity, the major emphasis is directed toward higher plants but algal and bacterial photosynthesis are also considered. I have attempted to deal with the information in an integrated fashion so that it is possible to view photosynthesis as a process sustaining and yet responsive to the general metabolism of the plant. Thus, the light reactions and electron transport are modified by the demands of the stroma for ATP and reducing power, while carbon metabolism, for example, is in turn regulated by sink requirements.

In dealing with the complexities of the biochemistry I have attempted to keep the text simple, comprehensive, and easy to follow, and I have tried to specify the areas in which current information is inconclusive. It is my hope that this treatise may provide a dependable source of information for advanced students, teachers, and research workers who are not expert in the field.

CHRISTINE H. FOYER

Sheffield, England
March 1984

ACKNOWLEDGMENTS

I am indebted to Dr. Richard Leegood for his advice, comments, and constructive discussions during the preparation of the manuscript. I gratefully acknowledge the support and expert comments provided by my colleagues at the Research Institute for Photosynthesis at Sheffield University and I thank Professor David Walker for leading me to a deeper understanding of the photosynthetic processes. I am most grateful to the following for the gift of photographs or figures or the permission to reproduce figures from their publications: Dr. Jan Anderson, Dr. Betril Andersson, and Elsevier Biomedical Press for Figure 2.4; Dr. Catherine Carver, Dr. Alex Hope, and the *Biochemical Journal* for Figure 4.4; Dr. Bill Cockburn for Plate 8.3; Dr. Ray Ellis for Plates 2.1 and 2.2 and Figure 2.1; Dr. Hal Hatch for Plates 7.2 and 7.3; Dr. Robert Hill and Dr. Peter Rich for Figure 2.3; Dr. Peter Horton and *FEBS Letters* for Figure 2.7; Dr. Anita Jellings for Figure 2.1; Dr. M. Kluge, Dr. I. C. Buchanan-Bollig, Dr. A. Fischer, and the American Society of Plant Physiologists for Figure 8.2; Professor Erwin Latzko and Gustav Fisher Verlag for Figure 6.3; Professor Rachel Leech for Plate 5.1, Figure 2.1, and with Blackie & Son Limited (Glasgow) for Figure 4.2; Dr. Denis Murphy and Springer-Verlag for Figure 6.5; Dr. Paul Quick for Figure 4.3; Professor David Walker for Figures 3.3, 4.4, 6.3, and 6.5; and Dr. Klaus Winter for Figure 8.3.

C. H. F.

CONTENTS

ABBREVIATIONS

ACP	Acyl carrier protein
ADP	Adenosine diphosphate
ALA	δ-Amino-levulinic acid
ATP	Adenosine triphosphate
BChl	Bacteriochlorophyll
CAM	Crassulacean acid metabolism
Chl	Chlorophyll
CF_0-CF_1	Coupling factor (ATPase)
C1-THFA	5,10-methylene tetrahydrofolic acid
CoA	Coenzyme A
Cyt	Cytochrome
DCMU	3-(3′,4′-Dichlorophenyl)-1,1-dimethylurea
DCPIP	2,6-dichlorophenolindophenol
DGDG	Digalactosyldiacylglycerol
DHAP	Dihydroxyacetone phosphate
DPGA	Diphosphoglycerate
ESR	Electron spin resonance
F6P	Fructose-6-phosphate
FBP	Fructose-1,6-bisphosphate

FBPase	Fructose-1,6-bisphosphatase
FCCP	Carbonyl cyanide-p-trifluoromethoxyphenylhydrazone
Fd	Ferredoxin
Fe–S	Iron–Sulfur center
FNR	Ferredoxin-NADP reductase
GAP	Glyceraldehyde-3-phosphate
GBP	Glycerate-1,3-bisphosphate
G1P	Glucose-1-phosphate
G6P	Glucose-6-phosphate
GOGAT	Glutamine (amide):2-oxoglutarate amino-transferase (NADP oxidoreductase)
GSH	Reduced glutathione
GSSG	Oxidized glutathione
LDS	Lithium dodecyl sulfate
LEM	Light-effect mediator
LHC	Light-harvesting chlorophyl-a/b (binding protein)
MGDG	Monogalactosyldiacylglycerol
OAA	Oxaloacetate
2OG	2-oxoglutarate
PAGE	Polyacrylamide gel electrophoresis
PC	Phosphatidyl choline
PEP	Phosphoenolpyruvate
PG	Phosphatidylglycerol
PGA	3-Phosphoglycerate
ph	phaeophytin
PSI	Photosystem I
PSII	Photosystem II
PQ	Plastoquinone
PMF	Proton motive force
R5P	Ribose-5-phosphate
RPP	Reductive pentose phosphate (pathway)
Ru5P	Ribulose-5-phosphate
RuBP	Ribulose-1,5-bisphosphate

S7P	Sedoheptulose-7-phosphate
SBP	Sedoheptulose-1,7-bisphosphate
SBPase	Sedoheptulose-1,7-bisphosphatase
SDS	Sodium dodecyl sulfate
$TMQH_2$	Tetramethylhydroquinone
UDP	Uridine diphosphate

PHOTOSYNTHESIS

1

GENERAL CONCEPTS

1.1.   INTRODUCTION

Green plants, algae, and certain bacteria are able to absorb the energy in sunlight and subsequently convert it into chemical energy, which they then store for future use as organic carbon compounds. The light-driven anabolism of carbon dioxide is termed *photosynthesis*, and the organisms that are able to carry out this process are true *photoautotrophs*.

In addition to light energy and carbon dioxide the photosynthetic process requires an oxidizable substrate. In all plants, algae, and the cyanobacteria, this substrate is water, and photosynthesis results in the formation of molecular oxygen, which is derived from this water by the removal of electrons and protons. The overall photosynthetic reaction in these photoautotrophs can be written as

$$n\mathrm{CO_2} + 2n\mathrm{H_2O^*} \xrightarrow[\text{enzymes}]{h\nu} n\mathrm{CH_2O} + n\mathrm{H_2O} + n\mathrm{O_2^*} \qquad (1.1)$$

where $\mathrm{CH_2O}$ is the empirical formula for carbohydrate synthesized as a result of photosynthetic activity.

Other bacteria are able to utilize sunlight to synthesize organic compounds but do not evolve oxygen. These photosynthetic bacteria are *chemoautotrophs*, as they require an additional substrate for photosynthesis, which is used in place of water to provide reducing equivalents for the light-driven reactions. The overall process in these bacteria might be written:

$$2\mathrm{AH_2} + n\mathrm{CO_2} \xrightarrow{\hspace{2cm}} 2\mathrm{A} + n\mathrm{CH_2O} + \mathrm{H_2O} \qquad (1.2)$$

where $\mathrm{AH_2}$ represents an oxidizable substrate such as hydrogen sulfide ($\mathrm{H_2S}$). The chemoautotrophic bacteria may be divided, for simplicity, into three main groups. The green sulfur bacteria, the Chloraceae (e.g., *Chlorobium thiosulfatophilum*) are strict anaerobes that oxidize hydrogen sulfide to sulfur during photosynthesis. The purple sulfur bacteria, the Thiorhodaceae, (e.g., *Chromatium vinosum*) are able to oxidize various sulfur compounds and certain organic compounds. The nonsulfur purple bacteria, the Athiorhodaceae (e.g., *Rhodospirillum rubrum*) are heterotrophic and primarily utilize organic compounds (particularly tricarboxylic acid cycle intermediates) either photosynthetically or in aerobic respiration in the dark. The requirement of an alternative substrate to water, as the primary electron donor for photosynthesis, clearly distinguishes the bacterial photosynthetic process in chemoautotrophs from plant photosynthesis. In bacterial photosynthesis the added substrate molecule (e.g., $\mathrm{H_2S}$) is oxidized in place of water, the substrate from which oxygen is derived in all plants, algae, and cyanobacteria.

In all photosynthetic organisms the overall photosynthetic process is found to consist of the following essential events:

1. *Light Absorption.* Light is absorbed by pigment molecules associated with protein complexes that are embedded in specialized lipoprotein membranes (Junge, 1977). The light energy is transferred to a reaction center that contains a type of chlorophyll-*a* molecule in a special environment.

2. *Charge Separation.* The absorption of a light quantum by the reaction center chlorophyll molecule results in excitation and the loss of an electron to an adjacent acceptor molecule and charge separation is initiated. Charge separation is the first chemical change in the complex reaction sequence of photosynthesis (Sauer, 1979).

3. *Electron Transport.* The loss of an electron from the reaction center pigment results in the formation of a strong oxidant. This promotes the donation of an electron from an oxidizable substrate, for example, water, to replace it. The electron derived from chlorophyll is removed from the site of origin by a series of vectorally arranged electron carriers and charge separation is established across the membrane. This facilitates the formation of ATP by a membrane-bound ATPase (Avron, 1981a). A unique exception to this is encountered in the chromatophores of the purple membranes of *Halobacterium halobium*, where the photosensitizing pigment is bacteriorhodopsin, a pigment related to the rhodopsin found in visual systems (see Section 1.5). This bacterium does not carry out electron transport but is able to synthesize ATP via a bacteriorhodopsin proton "pump" (Stoeckenius, 1979).

4. *Energy Storage.* The chemical energy obtained from sunlight by the above sequence of events is stabilized and consolidated in the synthesis of organic compounds from carbon dioxide (Robinson and Walker, 1981).

1.2. LIGHT ABSORPTION AND ENERGY TRANSFER

The ability to absorb, harness, and utilize the energy content of light is a primary requisite for photosynthesis. The processes of light absorption and energy conversion are carried out in specific pigment–protein complexes embedded in the hydrophobic environment of the photosynthetic membranes (Boardman et al., 1978; Thornber et al., 1979). In prokaryotic cells the photosynthetic "chromatophore" membranes are associated with the extremely complex infoldings of the cell membrane (Thornber et al., 1978). In eukaryotic organisms, the photosynthetic membranes are localized in specialized organelles of the plastid-type, which are called *chloroplasts* because the substantial inner membrane system is highly enriched in the photosynthetic pigment chlorophyll (Arntzen and Briantais, 1975). In the chloroplast, the photosynthetic membrane is a highly organized structure that is rather fluid in nature and generally folds into stacked regions, which were called *thylakoids*, by Menke (1962), connected by nonstacked stromal lamellae.

Light absorption is facilitated by the presence of photosensitizing pigments. The possession of at least one type of chlorophyll molecule appears to be virtually universal in photosynthetic organisms. The chlorophylls, of which there are at least eight kinds, have a similar molecular structure (Figure

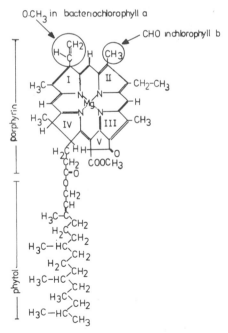

Figure 1.1. Diagrammatic representation of the structure of chlorophyll-*a*.

1.1) and are related biosynthetically and structurally to hemes. Like hemes, chlorophylls have a central tetrapyrrolic nucleus but, instead of iron, a molecule of magnesium is chelated within the center (Hill, 1963). In addition chlorophyll contains a pentanone ring and a phytol side chain, which is esterified to the C7 atom of the tetrapyrrole nucleus. Chlorophyll plays a central role in photosynthetic energy conversion. A small percentage of the total chlorophyll molecules (the reaction center chlorophylls) are bound in special environments where they form the "traps" for the absorbed energy and are the sites of the initial photochemical events.

Higher plants contain chlorophyll-a and chlorophyll-b while algae contain other chlorophylls (c,d,e) in addition to chlorophyll-a. Bacteriochlorophyll-a and bacteriochlorophyll-b are found in most photosynthetic bacteria. Most of the chlorophyll content of the photosynthetic membranes ($\sim$99%) serves to absorb light and channel excitation energy to the sites of photochemistry or "energy traps," in the reaction centers (Sauer, 1981). This "light-harvesting" function is also fulfilled by various other "accessory" pigments that supplement the light-gathering ability of chlorophyll. There are two major classes of nonchlorophyllous antenna pigments; these are the carotenoids and phycobiliproteins. Carotenoids, which occur universally in photosynthetic organisms (Britton, 1976), are long-chain polyunsaturated hydrocarbons with projecting methyl groups. The phycobiliproteins, which are abundant in cyanobacteria and red algae where they appear to replace chlorophyll-b, are conjugated proteins containing linear tetrapyrrols (Bogorad, 1975). The light-harvesting pigment molecules are organized into ordered "antenna systems." These consist of several types of pigment that absorb light maximally at different wavelengths so that virtually the whole of the visible spectrum can be exploited (Figure 1.2). Sunlight outside the earth's atmosphere has a total irradiance of 1353 W/m^2. However, passage through the atmosphere filters out all radiations below 300 nm and several regions in the infrared are virtually absent. At sea level approximately 75% of the total energy of sunlight is found in wavelengths between 400 and 1100 nm. It is within these limits that the photosynthetic pigments responsible for light capture have effective absorption. In higher plants the antenna pigments, chlorophyll-a, chlorophyll-b, and carotenoids, predominantly absorb light in the blue and red portions of the visible spectrum (Figure 1.3).

Radiant energy may be described in the form of discrete packets known as quanta. Each quantum or photon has an energy of $h\nu$ in which h is

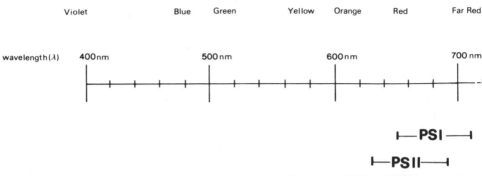

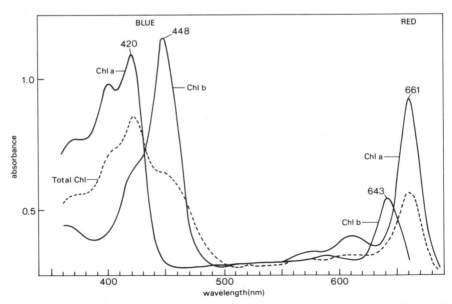

Figure 1.2. The visible spectrum, showing the absorption range of PSI and PSII in the red region.

Planck's constant (1.58×10^{-37} kcal · sec/photon) and ν is the frequency of radiation in sec^{-1}. Therefore the energy, E, of 1 mole (einstein) of photons is

$$E = N_A h\nu = N_A h \frac{c}{\lambda} = \frac{2.86 \times 10^4}{\lambda} \text{ kcal/einstein} \qquad (1.3)$$

where N_A is Avogadro's number (6.02×10^{23} photons/einstein), c is the velocity of light (3.0×10^{17} nm/sec), and λ is the wavelength in nanometers.

Figure 1.3. Absorption spectra of chlorophyll-*a* and chlorophyll-*b* and a total chlorophyll extract from *Spinacia oleracea* leaves in 80% acetone.

From equation (1.3) it can be seen that energy (E) is inversely proportional to wavelength and, therefore, photons at the violet end of the visible spectrum have the highest energy and photons in the far red region have the lowest energy. Light of 1200 nm and above has an energy content that is too low to mediate chemical change and energy absorbed in this range can only be converted to heat. Short wavelength radiation (<200 nm) gives rise to ionizing radiation such as X-rays, which have such a high energy content that they ionize the molecules that they encounter. Between 200 and 1200 nm the energy content is sufficient to produce a chemical change in an absorbing molecule. However, the chlorophyll molecule always relaxes to its lowest excited state before the light energy promotes this chemical change. Thus, although blue light has a high energy content that gives rise to the higher excited states of chlorophyll (Figure 1.4), it is no more effective than the less energetic red quanta (which generate the lowest excited state) in promoting the primary charge separation of photosynthesis.

Absorption of a light quantum by an antenna pigment such as chlorophyll

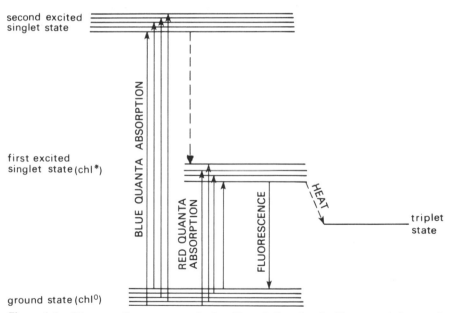

Figure 1.4. Diagram of energy states in the chlorophyll molecule. The ground, first, and second excited singlet states of the molecule possess a series of energy sublevels. Heat loss occurs when the excited molecule reverts to the lowest sublevel energy of the excited state. The fluorescence emission is shifted to the red end of the spectrum relative to the excitation spectrum because the light emitted in fluorescence is less than that absorbed during excitation (Stokes shift).

causes a transition or excitation of the molecule from the normal or ground state to an excited state. This excitation is rarely released (e.g., as fluorescence) and is normally transferred through the antenna pigment bed until it encounters an energy trap in a reaction center. In a pigment bed of identical molecules, for example, chlorophyll-*a*, homogeneous energy transfer between the molecules occurs. To achieve maximal efficiency the pigment molecules are oriented with respect to each other (Garab et al., 1981) so that they are no farther apart than 70 Å and no closer than 10 Å. The transfer of excitation energy between these adjacent chlorophyll molecules is not simply a process in which one molecule emits a quantum to be absorbed by the next. Within an array of like molecules in a chlorophyll protein complex an excited singlet state can become rapidly delocalized over the entire number of pigment molecules. This means that the internal interactions in the complex can be "excitonic" in nature. In this situation the excitation energy would be the communal property of all the like molecules, which may then be considered to act as a single large molecule or "exciton." However it is more likely that the excitation moves around the pigment bed from molecule to molecule in a random fashion until it is either lost by fluorescence or encounters a "trap." During this "energy migration" the excitation energy is considered to move between the pigment molecules by an inductive "resonance" process as described by Förster (1965). All heterogenous energy transfer between unlike molecules, for example, between chlorophyll-*b* and chlorophyll-*a*, occurs by the process of *resonance transfer*. When a pigment molecule absorbs a quantum of light the energy of that quantum elevates the molecule from its normal state of lowest energy and highest stability to an excited state of increased energy. In addition to this the molecule acquires vibrational energy resulting from this nonequilibrium state and thus the excited molecule vibrates or oscillates. Resonance transfer between molecules with overlapping absorption bands is a dipole–dipole interaction. The deexcitation oscillation in one molecule is coupled to a sympathetic oscillation in a neighboring molecule and the energy is directly transferred. Because of relaxation within the vibrational substates during the lifetime of the excited state, energy transfer between unlike molecules usually travels in one direction only.

1.3. CHARGE SEPARATION

Excitation energy is funneled toward the photochemical reaction centers in which specialized pigment molecules (usually chlorophyll) perform the energy conversion. When such a "quenching" reaction center is encountered the excitation energy is quickly abstracted from the antenna pigment system. The reaction center chlorophylls are often found as "special pairs," or dimers, a situation that aids energy delocalization. They are designated by the letter P for pigment and their wavelengths of maximum absorption. Thus we have P_{680} and P_{700} as the reaction center chlorophylls in photosystem II (PSII) and photosystem I (PSI), respectively, in higher plants and cyanobacteria. P_{870} and P_{840} are reaction center bacteriochlorophylls in chemoautotrophic bacteria. These are the "traps" for light energy and are arranged in a molecular complex in close proximity to other molecules, which are electron acceptors and electron donors to the reaction center chlorophyll. The ground state of these chlorophylls is generally the singlet state, that is, the molecules have only paired electrons of opposite spins (Figure 1.4) in the outermost orbit and thus have a total spin quantum number of 0. Absorption of light energy elevates one of the electrons from its closest permitted orbit around the nucleus or ground state (with energy E_0) into a wider orbit (with an energy of E_1) and thus the molecule is elevated to its excited state (Chl*). The overall state of the excited molecule is such that all of the spins are still paired even though the energy content is greater than that of the ground state. Within picoseconds of this energy transition, the electron in the wider orbit is captured by an adjacent acceptor molecule (A), which becomes reduced. The chlorophyll minus the electron is a strong oxidant that captures an electron of a lower energy from an adjacent donor molecule (Z) and returns to the ground state. The initial charge separation produces a "radical pair," in which the electron spins are initially antiparallel as in their singlet precursor. This series of events can be represented as follows:

$$Z \cdot Chl \cdot A \xrightarrow{h\nu} Z \cdot Chl^* \cdot A$$

$$Z \cdot Chl^*A \longrightarrow Z \cdot Chl^+A^- \qquad (1.4)$$

$$Z \cdot Chl^+A^- \longrightarrow Z^+ \cdot Chl \cdot A^-$$

The primary photochemical process is thus essentially an oxidation-reduction reaction and constitutes the first step of energy storage. The acceptor molecule (A) is generally found to be a dimeric chlorophyll as in PSI or a phaeophytin molecule as in the chemoautotrophic bacteria and PSII. The electron is rapidly transferred to a second electron carrier and thence through a chain

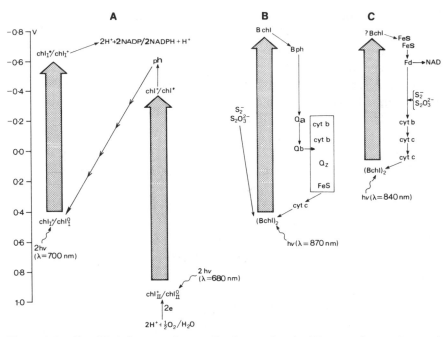

Figure 1.5. Simplified diagram of noncyclic electron flow in (A) oxygenic organisms and cyclic electron flow in (B) purple bacteria and (C) green bacteria.

of carriers that are situated vectorally across the membrane. In the chemoautotrophic bacteria electron transport is generally a cyclic process and electrons are introduced from an oxidizable substrate to establish an appropriate redox poise in the electron carriers (Figure 1.5). A side electron transport chain leads to the reduction of NAD to form NADH. In plants and algae cyclic electron transport also occurs but linear electron transport is the primary mode of photochemical action, leading directly to the reduction of nicotinamide adenine dinucleotide phosphate (NADP) to form NADPH. Water is the primary electron donor for this process, for each electron transferred from P_{680}, one electron is removed from water and with the removal of four electrons, one molecule of oxygen is released (Figure 1.5).

1.4. PHOTOOXIDATION

At low light intensities there is a high probability that most of the available quanta will excite a corresponding number of reaction centers. At high light intensities not all of the available quanta can be utilized to promote chemical change and superfluous light energy can be dissipated in several ways. Direct reemission of light can occur by the process of fluorescence or energy can be released by nonradiative transitions and lost as heat. Excess energy present in the excited singlet chlorophyll molecules must be rapidly dissipated in all aerobic organisms or the chlorophyll molecules themselves can be irreversibly bleached and destroyed in a harmful process known as photooxidation. Photodynamic oxygen activation occurs because during excitation some of the chlorophyll molecules attain triplet character. Chlorophyll in the triplet state can react directly with molecular oxygen and transferring energy raise normal triplet ground state oxygen to the much more reactive singlet state, $^1O_2$ ($'\Delta g$).

$$^3\text{chlorophyll} + O_2 \longrightarrow \text{chlorophyll} + {}^1O_2 \qquad (1.5)$$

Singlet oxygen has a relatively long lifetime in the hydrophobic environment of a membrane. In addition to chlorophyll destruction, singlet oxygen will rapidly react with the polyunsaturated fatty acid side chains in the membrane to form lipid peroxides. Lipid peroxidation is a highly destructive autocatalytic process that can lead to the loss of integrity and eventual fragmentation of the membrane. Prolonged illumination with high light is known to cause marked lipid peroxidation in thylakoid membranes. The carotenoid pigments, which have a light harvesting function, also serve to dissipate harmlessly the energy of the triplet chlorophyll molecule in the "triplet valve" mechanism (Witt, 1979). Carotenoid pigments such as β-carotene quench the triplet excited state of chlorophyll directly in a triplet–triplet resonance transfer interaction. The carotenoid molecule is not destroyed by photooxidation and carotenoids are efficient scavengers of singlet oxygen. The presence of carotenoid pigments is, therefore, a prerequisite for oxygenic photosynthesis. Mutants of maize that are deficient in carotenoids rapidly produce symptoms of photooxidation when illuminated; the chlorophyll is bleached and the thylakoid membranes show signs of fragmentation. The pathway of energy transfer from chlorophyll to carotenoid only occurs at a significant rate when the photosynthetic process is light saturated. Under such conditions approximately 20% of the superfluous energy can be dissipated by this means in chloroplasts. The pathway can also occur in photosynthetic bacteria but only approximately 1% of the excess excitation energy is dissipated in this way. In addition to the protection afforded by carotenoids, the thylakoid membranes in chloroplasts are safeguarded against harmful photooxidative damage by the presence of substantial amounts of α-tocopherol, which is a powerful singlet oxygen scavenger and an inhibitor of lipid peroxidation.

1.5. PHOTOPHOSPHORYLATION

The flow of electrons through the electron transport chain is coupled to the synthesis of the high energy phosphate adenosine triphosphate (ATP) (Arnon et al., 1954; Frenkel, 1954). The rate of ATP synthesis and the rate of electron transport are, therefore, interdependent. The method by which electron transport is coupled to and interacts with ATP synthesis is best explained by the chemiosmotic hypothetical model developed by Mitchell (1961, 1966), although other models may be considered (Table 1.1). The chemiosmotic hypothesis relies on the essentially vectoral arrangement of electron carriers within the photosynthetic membranes and their imperme-

TABLE 1.1.   HYPOTHESIS FOR COUPLING ATP SYNTHESIS TO
ELECTRON TRANSPORT

1.   The rate of electron transport is constrained by the rate at which energy can be used for ATP synthesis; this phenomenon is known as coupling. All coupling hypotheses utilize a high energy intermediate ($\sim$), which couples electron transport to ATP formation. Accumulation of ($\sim$) restrains the rate of electron transport. The action of uncouplers is to dissipate ($\sim$). The distinction between different coupling hypotheses lies in the nature of ($\sim$).

2.   In the *chemiosmotic hypothesis* the coupling intermediate ($\sim$) is a thermodynamic state achieved by a proton potential or proton motive force (PMF). The negative free energy associated with pumping protons down the ion gradient is coupled with the positive free energy change associated with ATP synthesis. The PMF is composed of two components, ΔpH, which reflects the difference in proton concentration on either side of the coupling membrane, and $\Delta\psi$, the membrane potential generated by the unidirectional transmembrane movement of charge.

3.   In the *chemical coupling hypothesis* electron transport generates a reactive chemical intermediate ($\sim$) that has sufficient energy to accommodate ATP synthesis. The coupling intermediate ($\sim$) is a molecule or complex with the required chemical properties to drive this and other energy-linked processes such as proton pumping.

4.   In the *conformational coupling hypothesis* transport of electrons elicits a conformational change in certain component macromolecules such that hydrolysis/ dehydration reactions, such as those involving ATP, are shifted in equilibrium to favor ATP synthesis.

ability to the passive flow of protons. The membranes thus have a distinct sidedness both functionally and physically. Illumination produces a charge separation across the membrane because the electron donors (e.g., P_{680} and P_{700}) and the electron acceptors (e.g., Q and X) are situated on opposite sides of the membrane. Junge and Witt (1968) were the first to provide evidence for the existence of an electric field or potential across the thylakoid membrane in the light. The activity of the light reactions in ejecting electrons across the membrane in a vectoral fashion simultaneously generates a localized field and an electric potential difference. At this stage the membrane is said to be energized (Witt, 1979).

The concept that ATP synthesis is "coupled" to electron transport is basic to the theory of photophosphorylation. Agents that stop ATP formation while allowing electron transport to proceed at maximal rates are called *uncouplers*. In terms of the chemiosmotic hypothesis the action of the uncoupler is to dissipate the proton gradient by permitting the passive movement of protons or other ions across the membrane. Thus the membrane is unable to establish a proton gradient and ATP synthesis is inhibited.

The membrane-bound enzyme complex that is responsible for the synthesis of ATP is called the *coupling factor* or *ATPase* (McCarty, 1979). The chemoautotrophic bacteria have an ATPase complex that is essentially analogous to the F_1–F_0 coupling factor of mitochondria. Chloroplasts too have a coupling factor that consists of two distinct components, a hydrophilic protein complex called CF_1 and a hydrophobic protein complex called CF_0. CF_1 projects into the external environment, for example, the stroma, and is the site of ATP synthesis while CF_0 is an intrinsic membrane protein concerned with the pumping of protons. CF_0 spans the membrane and forms the binding site for the hydrophilic CF_1 complex. The chloroplast coupling factor (CF_0–CF_1) is a reversible ATPase. It is complex both in its subunit composition and its substrate and effector binding properties. The CF_0–CF_1 complex is light activated and appears to be subject to close metabolic control *in vivo*, where it can exist in several states of differing catalytic activity. CF_1 is composed of five different polypeptides that are designated as α, β, γ, δ, and ϵ in order of decreasing molecular weight (Figure 1.6). The active site of the ATPase appears to be associated with the β subunits of which there are two per complex. The ϵ subunits (there are again two per complex) appear to inhibit the activity of the ATPase and it is possible that the complex is activated by a conformational change which moves the ϵ subunits away from the active site. The γ and δ subunits may have roles in the coupling

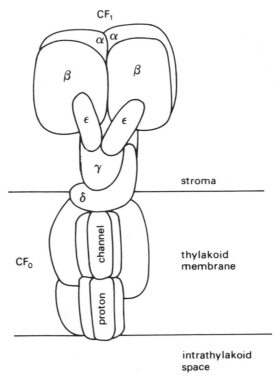

Figure 1.6. Diagrammatic representation of the coupling factor ATPase complex (CF_1-CF_0) in the thylakoid membrane.

of ATP synthesis and hydrolysis to proton fluxes. Jagendorf and Hind (1963) were the first to report a pH increase in the suspension medium of illuminated thylakoids. This alkalization is produced because for each of the light reactions in oxygenic photosynthesis one proton is taken from the outside to the inside of the membrane (Avron, 1981b). There are two sites of proton uptake on the outer surface of the thyakloid and two sites of proton release on the internal surface. According to the chemiosmotic hypothesis one proton is taken up from the outer surface of the thylakoid by the reduced plastoquinone pool and one by the proton-binding terminal acceptor, NADP. The oxidation of $\frac{1}{2}H_2O$ on the internal surface of the thylakoid releases a proton and the oxidation of the plastoquinone pool simultaneously liberates a proton into the intrathylakoid space. The number of protons transferred per electron traveling through the electron transport chain in oxygenic photosynthesis is generally considered to be $H^+/e = 2$, but may vary between 2 and 3.

This translocation of protons generates a ΔpH across the membrane of between 2.5 units at the membrane boundaries and 4.0 units in the bulk phase. The energization of the membrane produces a conformational change in the coupling factor thereby opening the gate of the proteolipid proton channel for the passage of protons and releasing the tightly bound adenine nucleotides, principally ADP. The electric field may simply produce this conformational change by the generation of a localized pH gradient across CF_1.

The synthesis of ATP is a condensation or dehydration reaction,

$$ADP + Pi \rightleftharpoons ATP + H_2O \qquad (1.6)$$

which can only proceed at a high rate when the water generated is quickly and completely removed. During photosynthesis the water that is a product of this reaction is not allowed to form as the components H^+ and OH^- are separated as they are removed from adenosine diphosphate (ADP) and inorganic phosphate (Pi). This is achieved by the pH gradient that is generated across the photosynthetic membrane in the light. According to the chemiosmotic hypothesis this ensures that there is an excess of H^+ on the inside of the membrane relative to the outside, which has an excess of OH^-. The synthesis of ATP continually neutralizes the proton gradient that is generated by electron transport.

An electrical potential $\Delta\psi$ is generated across the membrane as a result of charge separation. Both the electrical membrane potential and the proton gradient drive ATP synthesis by pumping protons through the proteolipid proton channel in CF_0. This causes a structural transformation in the CF_1–CF_0 complex, which converts CF_1 into an ATP synthesizing system. A membrane potential across the CF_1–CF_0 complex is effectively converted into a ΔpH across CF_1 because of the presence of the H^+-specific channel through CF_0. In dark adapted chloroplasts the CF_1–CF_0 complex is catalytically inactive but becomes active when ΔH^+ is impressed across the thylakoid membrane, either in the form of ΔpH or $\Delta\psi$. Together the proton gradient and ψ form the PMF:

$$PMF = \Delta\psi - Z\Delta pH \qquad (1.7)$$

where Z is equal to 2.303 RT/F and has a value of 59 mV at 25°C and 298 K. R is the gas constant, T is the absolute temperature, and F is the Faraday constant. Thus the sum of the chemical ($-Z\Delta pH$) and electrical ($\Delta\psi$) components of the gradient provides the driving force for photophosphorylation.

The ΔpH that has been found to exist across the photosynthetic membranes in chloroplasts in the light is approximately 3 pH units. In most cases the $\Delta\psi$ is largely neutralized by an influx of chloride ions and an efflux of magnesium ions and therefore the contribution made by ψ is small. However, the chromatophore membranes of the purple bacteria are much less permeable to the passive movement of ions such as Mg^{2+} and Cl^-. The buildup of the membrane potential in this situation is found to limit the inward pumping of protons and also drives H^+ out through the ATPase. In these circumstances $\Delta\psi$ makes a significant contribution to the PMF.

The photosynthetic purple membranes of the bacterium *Halobacterium halobium* contain a single pigment, bacteriorhodopsin, that functions as a proton pump to generate $\Delta\psi$ and drive ATP synthesis. The bacteriorhodopsin pump translocates a charged atom (H^+) and is thus coupled to the generation of an electric field. The energy to drive the pump is supplied by the photoreaction (Stoeckenius, 1979). The purple chromatophores or "patches," are not always present but develop on the cytoplasmic membrane in response to an environment where light is available but oxygen is scarce. Illumination generates a PMF across the chromatophores. Protons are withdrawn from the inside of the membrane and extruded from the outside. This process is accompanied by a cycle of conformational changes in the bacteriorhodopsin lattice. However *H. halobium* is not completely autotrophic and cannot survive solely by photosynthesis. The purple membranes are utilized to supplement the oxidative phosphorylation pathway in situations where oxygen is limiting.

1.6.   THE UTILIZATION OF ATP AND NADPH

The energy stored in the phosphate linkage of ATP and the reducing power in NADH or NADPH that are generated as a result of electron transport are utilized and consolidated in the dark reactions of photosynthesis. The energy temporarily stored in these compounds is used to drive the fixation of carbon dioxide into carbohydrate, via the reductive pentose phosphate (RPP) pathway, and to power all the biosynthetic reactions of the cell in the light.

ATP is a molecular carrier of free energy in a readily accessible form. It has two "high energy" phosphate anhydride bonds

$$ATP + H_2O \rightleftharpoons ADP + Pi + H^+ \qquad \Delta G_0' = -7.3 \text{ kcal/mole}$$

$$(1.8)$$

$$ADP + H_2O \rightleftharpoons AMP + Pi + H^+ \qquad \Delta G_0' = -7.3 \text{ kcal/mole}$$
$$(1.9)$$

Biosynthetic reactions are often thermodynamically unfavorable. ATP is used to phosphorylate intermediates, thereby producing activated derivatives that can react spontaneously. The energy is utilized by chemical coupling through an activated intermediate formed by transfer of a phosphate, pyrophosphate, or adenylate group to one of the reactants. Generally pyrophosphate or adenylate coupling is used when the biosynthetic reaction is very unfavorable or when complete irreversibility is essential. For example, the first committed step in sucrose biosynthesis is the formation of uridine diphosphate (UDP) glucose by UDP–glucose pyrophosphorylase [equation (1.10)] and that of starch biosynthesis is the formation of ADP glucose via ADP–glucose pyrophosphorylase [equation (1.11)]:

$$UTP + \text{glucose-1-P} \rightleftharpoons UDP \text{ glucose} + PPi \qquad (1.10)$$

$$ATP + \text{glucose-1-P} \rightleftharpoons ADP \text{ glucose} + PPi \qquad (1.11)$$

Although there is only a very small standard free energy change in these reactions the subsequent hydrolysis of the pyrophosphate (PPi) by efficient pyrophosphatase has the effect of making the reactions irreversible in the direction of synthesis of the sugar nucleotide with an overall $\Delta G_0'$ of about -7.0 kcal. ATP and ADP cycle between the photochemical energy generating reactions of the membranes and the energy requiring reactions of biosynthesis.

NADPH is a universal carrier of hydrogen and electrons. The nicotinamide ring of the pyridine nucleotide can accept two electrons and one H^+. When NADP is reduced as a result of electron transport a second H^+ is taken up. Therefore most redox reactions in biological systems take the form:

$$AH_2 + NAD(P)^+ \rightleftharpoons A + NAD(P)H + H^+ \qquad (1.12)$$

NADH and NADPH have different metabolic functions. NADH is generally used in ATP synthesis by oxidative phosphorylation and transfers $2H^+$ and $2e^-$ to oxygen with concomitant ATP production, while NADPH transfers $2H^+$ and $2e^-$ to oxidized precursors in the reduction reactions of biosynthesis. NADPH cycles between the photosynthetic membranes and the biosynthetic reactions, carrying reducing power, in a similar manner to ATP, which serves as the carrier of free energy.

In general the pathways for the breakdown and biosynthesis of a particular metabolite are distinct, utilizing one or more unique enzymes. The first unique enzyme of a biosynthetic pathway is generally a regulatory enzyme and is often directly or indirectly responsive to the concentration of ATP. Distinct pathways for carbohydrate biosynthesis are necessary to allow degradation or synthesis to proceed with a net negative $\Delta G_0'$. Separate or unique enzymes permit independent control of degradation and synthesis according to the requirements of the cell.

REFERENCES

Arnon, D. I., Allen, M. B., and Whatley, F. K. (1954). *Nature* **174**, 394–396.

Arntzen, C. J. and Briantais, J-M. (1975). In *Bioenergetics of Photosynthesis* (Govindjee, ed.), pp. 51–113. Academic Press, New York.

Avron, M. (1981a). In *The Biochemistry of Plants* (M. D. Hatch and N. K. Boardman, eds.), Vol. 8, pp. 164–193. Academic Press, New York.

Avron, M. (1981b). In *Photosynthesis* (G. Akoyunoglou, ed.), Vol. II, Photosynthetic Electron Transport and Photophosphorylation, pp. 917–928. Balaban International Science Services, Philadelphia.

Boardman, N. K., Anderson, J. M., and Goodchild, D. J. (1978). *Current Topics in Bioenergetics* **8**, 35–103.

Bogorad, L. (1975). *Ann. Rev. Plant Physiol.* **26**, 369–401.

Britton, G. (1976). In *Chemistry and Biochemistry of Plant Pigments* (T. W. Goodwin, ed.), pp. 262–327. Academic Press, New York.

Förster, T. (1965). In *Modern Quantum Chemistry* (O. Sinanöglu, ed.), Part III, pp. 93–137. Academic Press, New York.

Frenkel, A. W. (1954). *J. Am. Chem. Soc.* **76**, 5568–5572.

Garab, G. I., Kiss, J. G., Mustárdy, L., and Michel-Villaz, M. (1981). *Biophys. J.* **34**, 423–437.

Hill, R. (1963). In *Comprehensive Biochemistry* (M. Florkin and G. H. Stotz, eds.), Vol. 9, pp. 73–97. Elsevier, London.

Jagendorf, A. T. and Hind, G. (1963). In *Photosynthetic Mechanism of Green Plants*, pp. 599–610. National Academy of Science Research Council, Washington, D.C.

Junge, W. (1977). In *Encyclopedia of Plant Physiology* (A. Trebst and M. Avron, eds.), New Series, Vol. 5, pp. 59–93. Springer-Verlag, Berlin.

Junge, W. and Witt, H. T. (1968). *Z. Naturforsch.* **23b**, 244–254.

McCarty, R. E. (1979). *Ann. Rev. Plant Physiol.* **30**, 79–104.

Menke, W. (1962). *Ann. Rev. Plant Physiol.* **13**, 27–44.

Mitchell, P. D. (1961). *Nature* **191**, 144–148.

Mitchell, P. D. (1966). *Biol. Rev.* **41**, 445–502.

Robinson, S. P. and Walker, D. A. (1981). In *The Biochemistry of Plants* (M. D. Hatch and N. K. Boardman, eds.), Vol. 8, pp. 194–237. Academic Press, New York.

Sauer, K. (1979). *Ann. Rev. Phys. Chem.* **30**, 155–178.

Sauer, K. (1981). In *Photosynthesis* (G. Akoyunoglu, ed.), Vol. III, Structure and Molecular Organization of the Photosynthetic Membrane, pp. 685–700. Balaban International Science Services, Philadelphia.

Stoeckenius, W. (1979). In *Membrane Transduction Mechanisms* (R. A. Cone and J. Dowling, eds.), Society of General Physiologists Series, Vol. 23, pp. 39–47. Raven, New York.

Thornber, J. P., Trosper, T. L., and Strouse, G. E. (1978). In *The Photosynthetic Bacteria* (W. R. Sistrom and R. K. Clayton, eds.), pp. 133–160. Plenum Press, New York.

Thornber, J. P., Markwell, J. P., and Reiman, S. (1979). *Photochem. Photobiol.* **29**, 1205–1216.

Witt, H. T. (1979). *Biochim. Biophys. Acta* **505**, 355–427.

2

THE PHOTOSYNTHETIC
MEMBRANES

2.1. THE CHLOROPLAST

In eucaryotic cells the photosynthetic processes are compartmentalized in discrete, highly specialized organelles called plastids (Kirk and Tilney-Basset, 1978). The photosynthetic plastids of algal cells are found in a variety of colors because of the presence of large amounts of variously colored antennae pigments (e.g., fucoxanthin, phycoerythrin, phycocyanin) in addition to chlorophyll-*a*. These plastids are given the general name *chromoplasts*. In some algae and all higher plants chlorophyll (*a* and *b*) is the predominant photosynthetic pigment and the green chlorophyll-containing plastids are called *chloroplasts*. Many algal species (e.g., *Chlorella, Spirogyra*) possess only a single large chloroplast per cell, which is located peripherally. Other algae and most higher plants are found to contain smaller plastids in large numbers, a situation that may be advantageous because of the increased surface area of the total chloroplast compartment in relation to its volume. The higher plant chloroplast is generally lens-shaped and ranges in size from 3–10 μm in length and 1–5 μm in diameter. The chloroplast is, however, a flexible organelle able to change in volume, shape, and position in response to a stimulus such as light.

The leaves of dicotyledonous plants from temperate zones generally have two chloroplast containing tissue layers, an upper palisade mesophyll and a lower spongy mesophyll. In these cells the chloroplasts are located peripherally forming a monolayer against the cell wall (Plate 2.1). The surface area of the cell that is covered by the chloroplasts appears to be characteristic and constant for a given species. For example, in mature mesophyll cells of wheat approximately 70% of the surface area of the cell is covered by chloroplasts. This spatial arrangement of the chloroplasts optimizes the interception of light and the availability of CO_2 diffusing from the intercellular air spaces. These structural features of mesophyll cells probably account for the observation that in a range of environmental conditions, the photosynthetic rate of leaves under saturating light is directly proportional to the surface area of the cells.

Protoplasmic streaming in higher plant cells causes movement of the cellular organelles in the peripheral protoplasmic layer, but this has little effect on the position of the chloroplasts. However, the chloroplasts of some species are able to reorient their position so that light absorption may be limited under excessive irradiation thus avoiding the harmful effects of photooxidation and photoinhibition. Similarly when light intensity is limiting

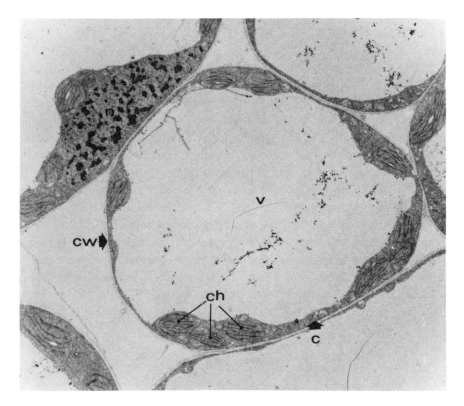

Plate 2.1. Developing mesophyll cells of the first leaf of wheat (*Triticum aestivum* var. Maris Dove) × 4760, showing cell wall (cw), chloroplasts (ch), cytoplasm (c), nucleus (n), and vacuole (v).

for photosynthesis the chloroplasts of some photosynthetic organisms are able to gather at those regions of the cell where light exposure is maximum. For example the single flat chloroplast of the green alga *Mougeotia* is oriented in response to light intensity and quality (Haupt, 1982).

Chloroplasts arise in the cell from preexisting plastids (Possingham, 1980). These, together with their specific genetic information, are tranferred to newly formed cells during both vegetative and sexual reproduction. During growth and differentiation of young cells multiplication of chloroplast numbers occurs by chloroplast division. This process is a type of binary fission in which a constriction develops in the center of the plastid. The constriction increases until the chloroplast is dumbbell-shaped and division is concluded by pinching off in the central region until the two daughter plastids separate

(Leech et al., 1981). On reaching maturity chloroplasts in higher plant cells occupy approximately 75% of the volume of the protoplasm (Figure 2.1).

The chloroplast is bounded by two topologically and functionally distinct membranes that are separated from each other by a gap of 10–20 nm, which is freely permeable to solutes such as sucrose and sorbitol and hence is sometimes called the *sucrose permeable space*. Together the two bounding membranes form the chloroplast envelope. The inner envelope membrane is occasionally found to proliferate inward forming folds that invaginate into the aqueous compartment of the chloroplast, the stroma.

The proliferation of the inner membrane can be extensive in plants that exhibit C4 photosynthesis (see Chapter 7). In the bundle sheath and mesophyll cells of these plants the extensions of the inner membrane form a system of anastomosing tubules called the *peripheral reticulum*. This structure greatly increases the surface area of the inner envelope membrane that is in contact with the stroma. The aqueous stromal compartment is highly viscous because of its substantial protein content (Table 2.1), however,

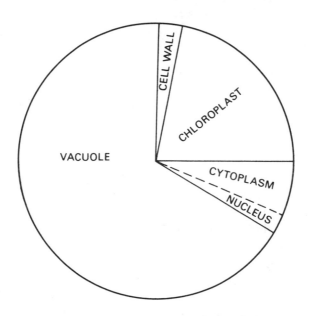

Total cell volume (average) $= 9052$ μm^3

Figure 2.1. The relative volumes of the subcellular compartments of mesophyll cells from *Asparagus officinalis*. The vacuole occupies approximately 70% of the cell volume, chloroplasts 21%, and cytoplasm 6%.

TABLE 2.1. CALCULATIONS OF
CHLOROPLAST DIMENSIONS

1 g of spinach leaf contains approximately
   400,000,000 chloroplasts
   1 mg of chlorophyll
   18 mg of chloroplast protein $\left\{ \begin{array}{l} 9 \text{ mg in stroma} \\ 9 \text{ mg in thylakoids} \end{array} \right.$

The chloroplasts occupy
   ~ 20% of the cell volume.
In 1 g of spinach leaf
   the total surface area of the thylakoids ~ 60 m²
   the total surface area of the envelope ~ 400 cm²
   the total volume of the chloroplast ~ 30 μL

Wildman et al. (1966) have shown that the stroma is in constant motion in the chloroplasts. It contains a vast array of synthetic enzymes as well as those of the CO_2-fixation pathway. Large starch grains are often present and there are usually small spherical osmiophilic bodies rich in lipids called *plastoglobuli*. These bodies vary from 10–500 nm in diameter and are embedded in the stroma between the membrane stacks but never within the grana or close to the envelope. They are generally considered to be reservoirs for excess lipid and contain compounds such as plastoquinone and tocopherylquinones. The stroma is rich in low molecular weight compounds and contains several copies of a covalently linked circular DNA molecule of molecular weight 85,000,000. Among the genes present on the chloroplast DNA are those coding for the chloroplast ribosomal RNA. Chloroplast ribosomes that reside in the stroma are believed to synthesize about 100 different polypeptides coded by chloroplast genes.

    The viscous protein matrix of the stroma surrounds and supports the internal photosynthetic lamellar membrane that has a unique structural organization. The lamellae or thylakoids, generally take the form of a complex network of closely appressed stacked membranes, the grana lamellae, connected by single unstacked membranes, the stroma lamellae (Plate 2.2). A grana stack consists of two or more membranes. The size and number of grana varies greatly. Not all photosynthetic cells show grana formation in the chloroplasts that may contain only stroma lamellae. The inner and outer surface of the membranes are clearly distinct. The outer surface is immersed in the stroma while the inner surface encloses an internal aqueous compartment, the intrathylakoid space, which is continuous between stacked

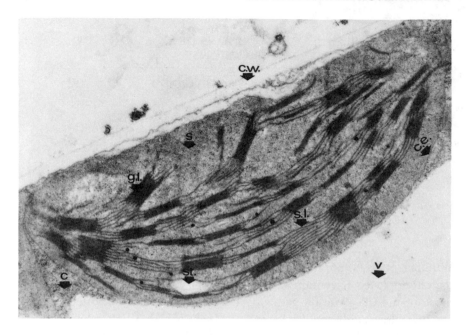

Plate 2.2.   Mature chloroplast of the first leaf of wheat (*Triticum aestivum* var. Maris Dove) ×31700 showing cell wall (cw), grana lamellae (gl), starch (st), stroma (s), stroma lamellae (sl), and vacuole (v).

and unstacked regions. It has been suggested that the thylakoid membranes are formed by invagination of the inner envelope membrane but the evidence for this is circumstantial (Kirk and Tilney-Basset, 1978). Direct connections between the plasmalemma and the outer thylakoid membranes have been observed in the prokaryotic cyanobacteria but these are rare. The protein, lipid, and pigment composition of the thylakoid membranes is qualitatively and quantitatively very different from that of the inner envelope membrane. The chloroplast membranes are rich in galactolipids and phospholipids. In the thylakoid membrane the two main galactolipds, monogalactosyldiacyl-glycerol (MGDG) and digalactosyldiacylglycerol (DGDG) are present in a ratio of 2:1. The two main phospholipids, phosphatidylcholine (PC) and phosphatidylglycerol (PG) are present in the thylakoids in a ratio of 1:3. In contrast the envelope contains MGDG and DGDG in a ratio of $0.3-0.8:1$ and PC and PG in a ratio of 3:1 (Douce and Joyard, 1979).

2.2.   THE REQUIREMENT FOR TWO PHOTOCHEMICAL REACTIONS

The overall photochemical reaction that is characteristic of oxygenic photosynthesis is a noncyclic process leading to the formation of ATP and $NADPH_2$:

$$NADP + H_2O + ADP + Pi \xrightarrow{\;h\nu\;} NADPH + H^+ + ATP + \tfrac{1}{2}O_2$$

$$(2.1)$$

The reduction of NADP (redox potential $Eo'pH\ 7 = -0.34$ V) by water ($Eo'pH = +0.8$ V) is very unfavorable thermodynamically as a dark reaction. The amount of light energy required to bring the reduction to equilibrium is at least 230 kJ/mole. In addition, 50 kJ/mole are required for the synthesis of ATP. Plants and algae show a requirement of 4 light quanta for each NADP reduced and 8–10 quanta are needed per O_2 evolved. This is satisfied by the presence of two distinct photochemical reaction centers that act in series so that two light quanta cooperate in the transport of each electron to NADP. Each reaction center has 300–400 light-harvesting chlorophyll-*a* molecules associated with it.

Emerson and coworkers were the first to provide evidence for the involvement of two different light reactions. In 1943 Emerson and Lewis observed that when the green algae *Chlorella* was illuminated with wavelengths in the far-red region (e.g., above 680 nm) there was a significant decline in the quantum yield of photosynthesis, although light in spectral range is absorbed by chlorophyll-*a*. This phenomenon is known as the *red drop effect*. In 1957 Emerson and coworkers discovered that the quantum efficiency could be enhanced by simultaneous illumination with quanta in the orange/red region and the far-red region (Emerson, 1958). Illumination with two wavelengths together give a higher rate of oxygen evolution than the sum of the two rates measured separately. This "enhancement" became known as the *Emerson enhancement effect* and arises when the single wavelength applied does not equally excite both photosystems. Both photochemical reactions are required to drive NADP reduction from H_2O, neither photoact is sufficient by itself. When light is absorbed only by the pigments of the first photoreaction center (PSI) as happens at 690 nm and above, the efficiency of energy conversion decreases. The efficiency of the light absorbed at these wavelengths is enhanced by the addition of light absorbed by both photoreaction center pigments and by the light-harvesting pigments (chlorophyll-*b*). Wavelengths below 690 nm can promote both photoreactions but they preferentially drive the second photochemical reaction (PSII).

2.3. THE ELECTRON TRANSPORT CHAIN

The photosynthetic light reactions consist of a series of steps that require the concerted interaction of a number of electron carriers and enzymes in addition to the pigments that sensitize the reactions. The electron transport chain of algae and plants that have two photosystems is considerably more complex than that of the chemoautotrophic bacteria that have a single photosystem. Electrons are transferred across a wider range of redox potentials in oxygenic photosynthesis because of the cooperation of the two photosystems that facilitate the elevation of electrons from a redox state capable of oxidizing water to a redox state sufficiently negative to reduce NADP. PSII and PSI are in communication with each other via a series of electron carriers, which is often represented by the Z-scheme model (Figures 1.5 and 2.2). Hill and Bendall (1960) proposed that the two photosystems were arranged in series within the electron transfer chain and using the form of an energy diagram the concept constitutes the Z scheme. Hill and Rich

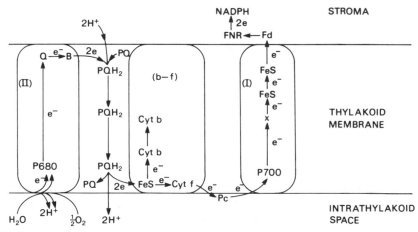

Figure 2.2.   The path of noncyclic electron flow drawn according to the Z scheme model, including the main protein complexes involved in electron transport; these are the cytochrome *b–f* complex (*b–f*) the PSII chlorophyll–protein complex (II) and the PSI chlorophyll protein complex (I).

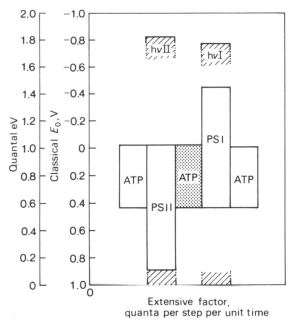

Figure 2.3. Modified diagram of the Z scheme. Taken from Hill, R. and Rich, P. R. (1983). *Proc. Natl. Acad. Sci. USA* 80, 978–982.

(1983) have provided a modified version of the scheme, in which the x axis is divided into five equal steps (Figure 2.3). The Z scheme is an excellent working hypothesis.

The proportion of electron transport chain components relative to the number of reaction centers is variable and dependent on growth conditions. The electron transport carriers are organized in a nonrandom asymmetric array across the membrane, for example, the coupling factor and NADP–ferredoxin reductase are located only on the outer surfaces of nonappressed membranes. The thylakoids have, in addition, a structural compartmentation of stacked and unstacked regions in which the photosystems have a distinct nonrandom distribution. In contrast to the photosystems, the complete chains of electron carriers do not have a physical reality. The two photosystems are discrete with separate pigment systems and are recognizable as physically separate particles of different size in freeze fracture studies under the electron microscope. The photosystems do not occur in a strict 1:1 proportion and measurements of photochemical activities show that they are separated in the lateral plane of the membrane. Anderson and Andersson (1981) have shown that there is an unequal distribution of the chlorophyll protein complexes in the appressed and nonappressed regions of the membrane. The stroma lamellae and the nonappressed regions of the membrane, which contain about 30% of the total chlorophyll, have been shown to contain most of the PSI complexes and only 10–20% of the PSII complexes. In contrast, the appressed regions of the grana stacks were found to be enriched in the PSII and light-harvesting complexes and substantially depleted of PSI complexes. Anderson and Andersson (1981) therefore suggest that the PSI complexes are localized only in the nonappressed regions of the membrane (Figure 2.4). The cytochrome b–f complex may be unevenly distributed between the grana and stroma lamellae.

If the photosystems are physically separated in this manner then one or more mobile electron carriers are required to carry reducing equivalents from PSII in the appressed grana membranes to PSI located in the nonappressed membranes. Two of the electron transport chain components, namely, plastoquinone and plastocyanin have what Bendall (1977) called a *distributive* function. These components could interconnect different chains and have only brief relationships with any specific reaction center and diffuse freely between them. The plastoquinone pool spans the membrane and transfers electrons from the primary electron acceptor of PSII, the quinone quencher Q. to the Rieske iron sulfur center, which is part of the cytochrome b–f complex. Plastoquinone molecules are lipid soluble and present in a greater concentration than the other electron carriers (Amesz, 1977). Amphiphatic molecules such as plastoquinone have fast diffusion rates in the lateral plane of the membrane and thus are highly mobile. In addition the reoxidation of reduced plastoquinone by P_{700}^+ is the rate limiting step in electron transport. The plastoquinone pool transfers protons across the membrane simultaneously with electron transport and thus contributes to ΔpH formation. Plastocyanin is water soluble, resides on the internal surface of the membranes, and may freely traverse the intrathylakoid space and thus interconnect appressed and nonappressed membranes (Katoh, 1977).

2.4. PSII AND OXYGEN EVOLUTION

The primary photochemical reaction of PSII reduces phaeophytin, a high potential electron acceptor, by transferring an electron from a very high potential donor, thus generating a strong reductant and a strong oxidant.

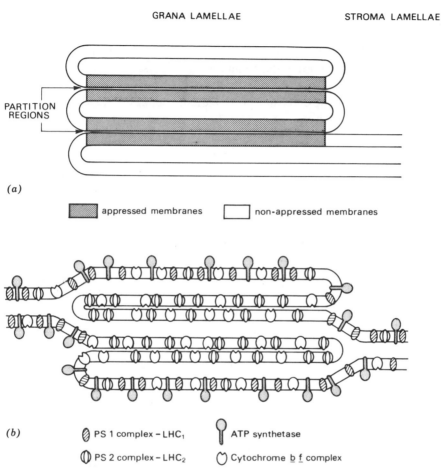

GRANA LAMELLAE   STROMA LAMELLAE

PARTITION
REGIONS

(a)

▨ appressed membranes   ☐ non-appressed membranes

(b)   🖋 PS 1 complex – LHC₁   🍾 ATP synthetase

        🔘 PS 2 complex – LHC₂   ⟆ Cytochrome b f complex

Figure 2.4. Diagrams to show the distinction between (a) appressed and stroma-exposed thylakoids and (b) the lateral distribution of the major complexes. (Taken from Anderson, J. M. and Andersson, B. (1982). *Trend. Biochem. Sci.* 7, 288–292.)

The reaction center chlorophyll and primary electron donor of PSII, P_{680}, can be observed at 680 or 820 nm. It has been proposed that P_{680} is a liganded monomeric chlorophyll-*a* (a situation that would render its electrochemical potential more positive than it would be if it were aggregated) or that it is a chlorophyll dimer. The similarity between the reaction center chlorophylls and their neighboring pigment molecules ensures good spectral overlap. This is necessary for efficient energy transfer and trapping in the reaction center. The primary electron acceptor in PSII has a midpoint potential of −610 mV and has been identified as pheophytin, a molecule that should be more readily reduced than chlorophyll. The product of the primary reaction is the radical pair ($P_{680}+$ pheophytin$^-$). The separated electrons on $P_{680}+$ and pheophytin$^-$ do not undergo an immediate loss of the spin correlation that they had in the excited singlet state (P_{680*} pheophytin) preceding the electron transfer. The electron spins gradually dephase because the electron on $P_{680}+$ and pheophytin$^-$ are in different environments and within pico-seconds the radical pair develops (P_{680+} pheophytin$^-$). Photosynthetic charge separation involving electron transfer can produce a change in the para-magnetism of the donor and acceptor molecules. At low temperatures, light induced free radical signals from many donor and acceptor species can be detected by electron spin resonance (ESR). P_{680} produces an ESR signal at $g = 2.003$.

The excitation of P_{680} primarily reduces the metastable acceptor pheo-phytin, which rapidly transfers the electron to the first stable acceptor to PSII called Q (for quencher), a semiquinone. $P_{680}+$ is re-reduced by a secondary donor that is then reduced by an electron derived from the oxygen evolving system. Each oxygen evolving complex is associated solely with one PSII reaction center. An electron donor to P_{680}, Z, can also be observed by ESR. Z produces a signal with a very rapid rise time and is, therefore, called signal $II_{very\ fast}$. This electron donor is probably also responsible for a slower ESR signal called signal II_{fast}. A number of signal II components with various decay times are, therefore, observable by ESR at room tem-perature. The signal II species, signal II_{fast}, appears at cryogenic temperatures suggesting that this species is close to the PSII reaction center (Nugent and Evans, 1979). Two electron carriers that are close to P_{680} but do not appear to be directly involved with oxygen evolution are cytochrome b-559 and, F, which give rise to a signal with a slow rise time called signal II_{slow} (Bouges-Bocquet, 1980). High potential cytochrome b-559 is closely as-sociated with Z, and with the manganese of the water-splitting system as all treatments extracting manganese lower the potential of this cytochrome. Optical changes at 320 nm have been reported to arise from a component close to the water-splitting site.

It is generally accepted that the bound manganese (5–8 Mn/400 chlo-rophylls) is an essential cofactor for the water-splitting system (Radmer and Cheniae, 1977). Some models of the oxygen evolving mechanism suggest that there are as few as two manganese per water-splitting "enzyme." In the thylakoids, manganese exists in a mixture of oxidation states, that is, Mn^{2+} and at least one higher oxidation state (Govindjee and Wydrzynski,

1981). Approximately two-thirds of the manganese is only loosely bound to the membrane. This is not detectable by ESR at room temperature and is at least in part directly involved with oxygen evolution. The remainder of the manganese is either tightly or very loosely bound and some of this may be associated with the light harvesting chlorophyll-a/b binding protein.

The mechanism of oxidation of water to molecular O_2, which can be represented as

$$2H_2O \longrightarrow 4e^- + 4H^+ + O_2 \qquad (2.2)$$

remains unsolved. Reaction (2.2) does not occur in a single step (Renger, 1978). Four positive charges, generated by the turnover of P_{680}, must accumulate before oxygen can be evolved. The generally accepted model of Kok and coworkers (Joliot and Kok, 1975) predicts that the water-splitting "enzyme" cycles through four oxidation states donated S, with a subscript that indicates the number of oxidizing equivalents, in excess of that on S_0, stored in the system (e.g., S_1, S_2). A subscript prime represents the immediate state created after light absorption by P_{680}. State 4, which has four oxidizing equivalents and is highly unstable, rapidly regenerates S_0 by evolving a molecule of oxygen:

$$S_0 \rightarrow S_0' \xrightarrow[H^+]{Z_1, e^-} S_1 \rightarrow S_1' \rightarrow S_2 \rightarrow S_2' \xrightarrow[H^+]{e^-} S_3 \rightarrow S_3' \xrightarrow[2H^+]{Z_2, e^-} S_4 \qquad (2.3)$$

$$O_2$$

This model is consistent with oxygen flash yield measurements. Joliot and coworkers (1968, 1969) using a rapidly responding sensitive oxygen electrode subjected dark adapted algae and chloroplasts to short saturating light flashes. Oxygen was evolved after the third flash followed by a damped oscillation of oxygen yields with a characteristic perodicity of 4 due to the cycling of the S states. The details of damping can be explained by the occurrence of misses, double hits, and that the ratio of $S_1:S_0$ in dark adapted chloroplasts is 3:1. "Misses" are randomly distributed failures of state transition or quantum absorption. "Double hits" occur when two successive state transitions occur within a single flash of light (i.e., 2 quanta absorbed). In the dark the oxidation states S_3 and S_2 are unstable and decay to yield S_1, the distribution is then approximately 25% S_0 and 75% S_1 with no S_2 or S_3. Double hits advance the phase of the cycle while misses retard it.

When thylakoids are washed with Tris buffer, manganese is liberated from the membranes and oxygen evolution is lost. Tris washing results in the release of ⅓ to ⅔ of the membrane bound manganese (4 atoms/trap) but this manganese may not all be directly derived from the oxygen evolving system. Several other treatments giving rise to similar results are high pH, heat treatment, and hydroxylamine treatment. Tris washing, high pH inactivation, and heat treatment have been shown to attack primarily the S_2 oxidation state. Analysis of the polypeptide composition and thylakoid bound manganese of mutants of the green alga *Scenedesmus obliquus* blocked specifically on the oxidizing side of PSII has shown that the loss of the larger loosely bound manganese pool is related specifically to the absence of a 34-kilodalton polypeptide.

Chloride ions appear to be essential to the functioning of the oxygen evolving system. In chloride depleted chloroplasts the ability to evolve oxygen is lost (Izawa et al., 1979). Chloride depletion also renders the system more susceptible to hydroxylamine attack or Tris inactivation and causes partial or total inactivation by added Mn^{2+}, which does not affect normal chloroplasts. These effects can be reversed by the addition of chloride ions suggesting the chloride depletion causes reversible modifications of the oxygen evolving system. Chloride is considered to act as a counterion for the charge accumulation in the water-splitting system. Depletion of the ion stabilizes the charge accumulation in states S_2 and S_3.

Bicarbonate stimulation of PSII in bicarbonate depleted chloroplasts is well-known. There is a major site of low affinity bicarbonate binding on the reducing side of PSII and a small high affinity binding site on the oxidizing side of PSII (Stemler, 1978). Catalytic amounts of CO_2/HCO_3^- are thought to bind temporarily to PSII in a cyclic fashion and it has been suggested that CO_2 may be directly involved in oxygen evolution. Bicarbonate is essential for the reincorporation of manganese into the thylakoid membranes after Tris treatment and will also relieve the inhibition of oxygen evolution by formate.

The structure of PSII is highly complex and can contain several types of primary and secondary electron acceptors and donors. The analysis of the kinetics of chlorophyll fluorescence has shown that thylakoids contain two distinct types of PSII centers that are called α and β. Both operate with high quantum efficiencies, evolve oxygen, reduce Q, and are associated with the C_{550} absorption change but they differ in the size of their antennae (Horton, 1982). Reduction of PSIIα causes a rapid sigmoidal fluorescence induction while the reduction of PSIIβ is slower with first-order reaction kinetics. The size of the PSIIα antenna pigment bed is large compared with PSIIβ and allows energy transfer between adjacent PSIIα units. This is not the case with the discrete smaller PSIIβ units. The ratio of PSIIα centers to PSIIβ centers is not constant. It is suggested that α centers are confined to the appressed membranes of the grana while β centers are found only in nonappressed regions. Redox titrations of the fluorescence yield have shown that in addition to the high potential acceptor Q_H or Q_1, which has a midpoint potential of $+10$ to -40 mV there is a low potential acceptor Q_L or Q_2 with an operating potential of -247 to -320 mV. It has been suggested that Q_H is directly involved with electron transport from P_{680} to plastoquinone while Q_L is perhaps involved in cyclic electron flow. A quinone binding protein B, which is the site of binding for the herbicide 3-(3′,4′-dichlorophenyl)-1,1-dimethylurea (DCMU), mediates electron transfer between the single electron carrier Q and the plastoquinone pool which is a two-electron carrier. Redox changes in the plastoquinone pool are intimately connected with redox changes in cytochrome b-563, which is a component of the b–f complex (Rich and Bendall, 1981). The b–f protein complex contains two molecules of cytochrome b-563 (-90 mV) (and perhaps a bound ferrodoxin NADP-reductase), a c-type cytochrome called cytochrome f ($+375$ mV) and a Reiske-type iron sulfur center ($+330$ mV). This protein complex transfers electrons from the plastoquinone pool to plastocyanin and thence to P_{700}.

2.5. PSI

PSI reduces a very low potential electron acceptor at the expense of a moderate potential electron donor. P_{700} is possibly an aggregate of two chlorophyll molecules, a "special pair." This structure leads to excitonic interaction that provides relatively low energy excited singlet states that aids efficient energy trapping. The "special pair" aggregate also has a decreased reduction or electrochemical potential relative to that of monomeric chlorophyll. At present the PS I reaction center is believed to consist of the primary electron donor, P_{700}, and four electron acceptors (Malkin, 1982). The primary electron acceptor to PSI is most probably also chlorophyll and is possibly also a dimer. This acceptor rapidly transfers electrons to an iron containing acceptor denoted X (-730 mV) (Ke et al., 1977). Two bound iron–sulphur centers A (-550 mV) and B (-590 mV) are the next components in the chain. X and the iron–sulphur centers A and B are visible using ESR (Evans and Cammack, 1975). The iron–sulphur center A transfers electrons to ferredoxin. A strict linear sequence of electron transfer has been considered to occur (component X → center B → center A). However, Nugent et al., 1981 have suggested that centers A and B function as equivalent electron acceptors from X. Ferredoxins are small water-soluble proteins that function as one electron carriers (Hall and Rao, 1977). Electrons are transferred from ferredoxin to NADP via the enzyme ferredoxin–NADP$^+$ reductase. This enzyme is situated on the stromal surface of the thylakoid, and takes single electrons from the ferredoxin pool to carry out the two electron reduction of NADP. Ferredoxin–NADP reductase is also able to recycle the electrons from reduced ferredoxin to the plastoquinone pool in the process of cyclic electron flow around PSI.

2.6. THE CHLOROPHYLL PROTEIN COMPLEXES OF PLANTS AND ALGAE

The high efficiency of energy transfer between the photosynthetic pigments is obtained by the accurate orientation of the pigments with respect to each other. In the 1960s two possible methods of pigment organization were considered to be possible in order to achieve the correct geometry. The first was that the pigments which are lipophillic were located in the lipid phase of the membrane where the lipid molecules could provide the required spacing and orientation. The second was that the chlorophyll and other pigments were associated with protein as is the situation with other porphyrins. Over the last few years the latter has been repeatedly shown to be correct. The photosynthetic pigments, in stoichiometric association with their specific apoproteins, have been extracted and purified from a wide variety of sources. Very few of the chlorophyll protein complexes are water soluble. The phycobiliproteins of cyanobacteria and red algae are located on specific complexes called *phycobilisomes*. These are located on the external surfaces of the photosynthetic membranes and are readily removed by washing with aqeuous buffers (Gantt et al., 1976). A bacteriochlorophyll protein that represents about 5% of the antenna pigments in green sulphur bacteria has also been extracted using aqueous buffers as has a light-harvesting peridinin-chlorophyll-*a* protein from dinoflagellates. In most cases the pigment complexes can only be extracted from the membrane by detergent solubilization followed by electrophoretic fractionation using polyacrylamide gel electrophoresis (PAGE). Current procedures fractionate at least two chlorophyll-*a* containing complexes, one associated with each photosystem plus three complexes containing chlorophyll-*a* and chlorophyll-*b* (Thornber and Markwell, 1981).

The light-harvesting phycobilin pigments of cyanobacteria and red algae are chemically bonded to protein and form complexes that are collectively called *phycobiliproteins*. In contrast the carotenoid and chlorophyll pigments of higher plants and green algae are not chemically bonded to their proteins. The association of the pigments with specific proteins has several structural and biochemical advantages. Spatial distances between antenna pigments and reaction center chlorophylls can be maintained to maximize energy transfer to the trap. Proteins can provide constant microenvironments in which the associated pigments are held. This allows the formation of the unique spectral species and can broaden the absorption spectrum. Chlorophyll-*a* is found *in situ* in a number of different spectral forms that reflect the different interactions of the pigment molecules with each other and with their apoproteins. The conformation and properties of protein can be regulated and thus the associations between different complexes can be modified as, for example, in the process of "spillover."

The photosynthetic pigment system is currently considered to consist of structurally and functionally distinct lipoprotein aggregates or complexes, each of which is composed of more than one type of polypeptide. The chlorophyll proteins of the PSI and PSII complexes contain up to 50% of the total pigment of the membrane, the remainder is associated with a pigment–protein complex whose function is light harvesting. In higher plants this protein contains chlorophyll-*a*, chlorophyll-*b*, and carotenoid and is called the *light-harvesting chlorophyll*-a/b (LHC) binding protein. In intact thylakoids these chlorophyll protein complexes extend across the lipid bilayer and can be observed by the technique of freeze fracture (Staehelin, 1981). Under the electron microscope, the supramolecular complexes appear as particles of 80–160 Å in diameter. The pattern of freeze fracture particles in the thylakoid membranes is unique both in particle density and particle distribution. The technique of freeze fracture is performed by cleaving the frozen granal membranes so that they split internally through their most hydrophobic regions. The half of the fractured membrane that was closest to the stroma is by convention designated the P face and the half closest to the intrathylakoid space or loculus is designated the E face. PF and EF, therefore, refer to the fracture faces and PS and ES, refer to the surfaces exposed by deep etching. The subscript s is used to donate the stacked grana regions while u is used to denote the unstacked stroma lamellae. Using the double-replica technique the EF_s and the PF_s have been shown to be complementary. The inner fracture face, EF_s, (the B face of Branton and Park, 1967) is a thin layer studded with large particles of about 17 nm in diameter that are occasionally seen in paracrystalline assay. The PF_s (the C face of Branton and Park, 1967) is thicker than the EF_s and contains many small particles. The large freeze fracture particules can only be detected in the stacked membrane regions. The stroma lamellae contain fewer particles than the grana lamellae and they are mostly of the smaller type (Staehelin et al., 1976). There is now considerable evidence to suggest that the largest particles on the EF_s are PSII complexes plus their associated LHC proteins. Some of the smaller particles on the PF_s and in the stroma lamellae are thought to be PSI complexes and their associated LHC proteins.

The composition of the PSI and PSII reaction centers or core complexes has been found to be rather uniform throughout the plants and algae. The PSI reaction center complex is a P_{700}-containing chlorophyll-a protein in which the chlorophyll-a:P_{700} ratio has been reported to be between 80:1 and 20:1, although a value of 40:1 is generally obtained. The PSI complex generally appears as a single polypeptide of 100,000–135,000 molecular weight. This polypeptide can be dissociated into two polypeptides of 60,000–70,000. The PSII–P_{680} unit appears to consist of three complexes, a core complex containing the reaction center P_{680} and two peripheral complexes. Two of these complexes are chlorophyll-a binding proteins and are associated with the reaction center itself, the third complex appears to be a light harvesting antenna protein serving PSII alone. Oxygen evolving PSII particles can be prepared; for example, Henry and Møller, 1981 have described a PSII vesicle containing four chlorophyll proteins: two chlorophyll-a binding proteins of molecular weights of 51,000 and 43,000 and two chlorophyll-a/b binding proteins at molecular weights 30,000 and 24,000. Four additional polypeptides with molecular weights of 33,000, 22,000, 19,000, and 18,000 were also present. The low temperature (77 K) fluorescence emission spectra of the vesicles showed a predominant peak at 695 nm and a very low emission at 729 nm. The 33 kD polypeptide is intimately associated with the O_2 evolving system and has been correlated indirectly with manganese binding at the water oxidation site. Other polypeptides which have been shown to localize at this site have molecular weight of 22–23 kD and 16–17 kD. The oxygen evolving sites appear to be situated in clefts on the inner surface of the thylakoid membranes. These clefts have dimensions of approximately 4 by 2·5 $Å^2$.

2.7. LHC BINDING PROTEIN

The thylakoid membranes carry a net negative charge that is predominantly derived from carboxyl groups associated with glutamic and aspartic acid residues residing on exposed segments of integral membrane proteins such as the coupling factor. The LHC binding protein complex is the major chlorophyll binding protein in the membranes of plants and green algae (Bennett, 1979a). Charged detergents such as sodium dodecyl sulfate (SDS) are frequently used to release or subfractionate the functional protein complexes that are present in the thylakoids as supramolecular complexes. When thylakoid membranes are heated in the presence of SDS the chlorophyll components are solubilized and the major LHC polypeptide is observed to have a molecular weight of 25,000 on PAGE. Frequently, a second LHC polypeptide is found to be present with a molecular weight of slightly less than the major polypeptide ($\sim$23,000). There are, therefore, two related LHC polypeptides. The 25,000 polypeptide is considered to be the apoprotein of the complex. Individual chlorophyll containing aggregates can be isolated by using conditions that cause less denaturation and dissociation or by employing anionic or zwitterionic detergents (e.g., *N*-laurylsarcosine). Under these conditions three complexes containing chlorophyll-*a* and chlorophyll-*b* are generally apparent on PAGE. Multiple banding of the chlorophyll-*a* and chlorophyll-*b* complexes on electrophoresis has been considered to be due to the separation of different oligomeric forms of the fastest migrating complex, the monomeric form of which was suggested to have a molecular weight of about 28,000. However, recent evidence has shown that there is probably more than one basic type of light-harvesting complex polypeptide, one individual type being specifically associated with each photosystem while the major form can associate with both PSI and PSII.

    LHC is absent from etioplasts and its formation during greening correlates with rapid synthesis of chlorophyll-*b* and grana formation. The protein(s) is encoded in the nucleus and is synthesized in the cytoplasm in a larger precursor form. This precursor(s) crosses the chloroplast envelope and is cleaved to a polypeptide of approximately 25,000. The polypeptide is then inserted into the thylakoid membrane and chlorophyll-*a* and chlorophyll-*b*

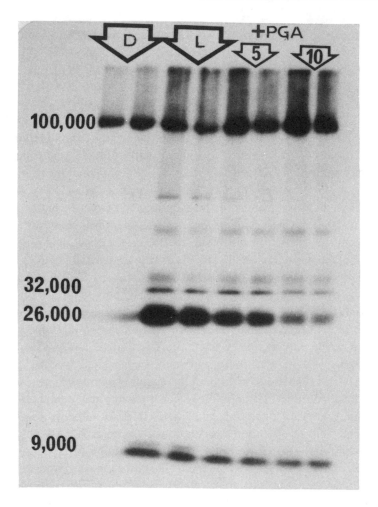

Figure 2.5. Autoradiogram made from an SDS-polyacrylamide slab gel of maize (*Zea mays*) mesophyll chloroplasts showing the polypeptides that were phosphorylated using ^{32}Pi for 5 min in darkness (D), 5 min in the light in the absence of substrate (L), and then with added PGA for 5 min and 10 min in the light.

are bound to it in a ratio of approximately 1:1. The LHC is a hydrophobic integral membrane protein but a small segment of 20 amino acids projects into the chloroplast stroma. Removal of this hydrophilic segment by trypsin digestion causes a loss of the stacking ability of thylakoids and also the loss of the ability of cations to regulate the direction of excitation energy transfer between the photosystems. Bennett and coworkers discovered that the LHC is phosphorylated by a membrane bound Mg^{2+} dependent, light-activated protein kinase (Figure 2.5) and dephosphorylated by a light-independent membrane bound phosphatase (Bennett, 1976b, 1980). The site of phosphorylation is a threonyl residue on the segment of the protein that is exposed to the stroma. The herbicide DCMU, which inhibits electron transfer from Q to plastoquinone, was found to inhibit the activation of the kinase. However, enzyme activity could be restored by the addition of reduced 2,6-dichlorophenolindophenol (DCPIP), which donates electrons to the electron transport chain at a position after plastoquinone. Ferricyanide and methyl viologen are efficient electron acceptors from PSI in isolated thylakoids and tend to cause oxidation of the electron transport carriers in the light. In the presence of these compounds phosphorylation was not observed in the light. Low potential reductants such as dithionite and tetramethylhydroquinone ($TMQH_2$) were found to activate the kinase and thus cause phosphorylation of the LHC in the dark. These observations suggested that the activation of the protein kinase that phosphorylates a number of polypeptides including the LHC protein was dictated by the redox state of one of the intersystem electron carriers. In 1981, Horton and coworkers provided evidence that the activation of the protein kinase was regulated by the redox state of plastoquinone.

2.8. SPILLOVER

The light-harvesting antenna pigment system in higher plants and algae is able to readjust the distribution of absorbed excitation energy between the two photosystems to maximize electron flow. Efficient noncyclic electron flow requires the balanced participation of both PSII and PSI. To accomplish this it is necessary to have equal distribution of excitation energy between the reaction centers of PSII and PSI. If one photosystem is excited more than the other, the excess excitation energy is dissipated and lost, and consequently the photosynthetic quantum efficiency decreases. However, under certain conditions where a large amount of ATP is required relative to NADPH, it may be advantageous for chloroplasts to carry out high rates of cyclic electron flow around PSI and thus drive cyclic photophosphorylation. In this situation an unequal distribution of excitation energy in favor of PSI would be beneficial. The light-harvesting system of the thylakoids allows this adaptive adjustment to be made by the process of *spillover*.

Bonaventura and Myers (1969) and Murata (1969) showed that the distribution of light energy of a fixed wavelength between PSII and PSI was variable. When subjected to continuous illumination with light that primarily excited PSII a slow change in the distribution of excitation energy occurred so that a better balance of PSII and PSI activity was achieved. The situation changed from one where maximal enhancement could be produced to one where both photosystems received equal excitation and therefore enhancement was minimal. The redistribution of excitation energy is an adaptive "transition," and was brought about by the spillover of excitation energy from PSII to PSI. State 2 refers to the adaptive state resulting from the over excitation of PSII and is characterized by increased energy transfer to PSI. Similarly, state 1 refers to a situation in which spillover of light energy is minimal and the capacity for enhancement is maximal. Adaptive changes between these states are thus called state 1–state 2 transitions (Figure 2.6).

Plants have developed these mechanisms as adaptations to changes in light quality; for example, species living on the ground below a dense forest canopy receive mainly light filtered through the leaves above which is often rich in far-red wavelengths. Similarly, algae living in the ocean depths receive light depleted in red and blue wavelengths but enriched in green. The mechanism by which the regulation of the photosynthetic light-harvesting properties is achieved has been the subject of numerous investigations. Plants that are adapted to growth in low light or shade have larger and more extensive grana stacks than have plants grown in high light (sun species). This has been suggested to indicate that grana formation is involved with an increase in the efficiency of light harvesting and excitation energy transfer to the photosystems. The ability to stack is generally lost when the environment of the membranes is depleted of cations. The membrane particles visible by freeze-fracture electron microscopy are found to be evenly distributed throughout the nonappressed regions in this situation. With the addition of cations grana formation is reestablished and the freeze-fracture particles are rearranged so that the large particles are found predominantly in the regions of membrane appression while the smaller particles are found mainly on the nonappressed membrane and stroma lamellae.

The concentration of cations in the thylakoid environment also has a profound effect on chlorophyll fluorescence. In the absence of divalent cations the association between the LHC binding protein complex and the photosystems is such that excitation energy is primarily transferred to PSI. Thus there is an increased physical association between the LHC and PSI

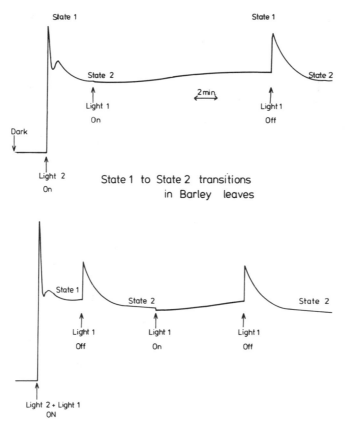

Figure 2.6. State 1–State 2 changes in a barley leaf at room temperature monitored via changes in modulated chlorophyll fluorescence. Utilizing a branched fiber optic, the leaf was exposed to a modulated light 2 (a blue-green PSII light at 1.3 W/m^2 transmitted by a Corning 4-96 filter) and a nonmodulated light 1, a far-red PSI light at 707 nm and 7.5 W/m^2.

in these circumstances. The addition of a 2–3-mM divalent cation concentration or a 100–150 mM monovalent cation concentration to a suspension of cation-depleted thylakoids is generally found to alter the distribution of excitation energy and mimic the changes that occur in state 1–state 2 transitions (Steinback et al., 1979). In limiting light, the addition of cations typically results in increased energy transfer to PSII, an increase in the quantum efficiency of PSII, increased variable fluorescence, and an increase in the 685/735-nm fluorescence at 77 K. Simultaneously the electron transfer to PSI declines. Thus the structural and functional relationships of the light-harvesting system are changed so that the transfer of excitation energy to PSII is increased.

The removal of the surface-exposed segment of the LHC by trypsin digestion causes the loss of both these cation effects. Similarly when specific antibodies to the LHC adhere to the protein *in situ* the cation effects are again lost. The facts strongly implicate the LHC itself in the regulation of the distribution of excitation energy between the photosystems. Allen and coworkers (1981) have proposed that the distribution of excitation energy is regulated via the phosphorylation of the segment of the LHC that is exposed to the stroma. The 77-K chlorophyll fluorescence emission at 735 nm is mainly derived from the chlorophyll-*a* molecules of PSI while that at 685 nm arises primarily from the light-harvesting chlorophyll of PSII. Phosphorylation of the LHC in isolated thylakoids was found to cause an increase in the emission at 735 nm relative to that at 685 nm, which suggested that phosphorylation caused a redistribution of excitation energy in favor of PSI relative to PSII. The protein kinase would be activated when the plastoquinone pool was reduced and thus it is probable that kinase activity will be modified in response to factors that will influence the redox state of plastoquinone. In addition to the effect of imbalance in the rates of excitation of PSII and PSI, kinase activity may change with light intensity, the size of the transmembrane pH gradient, and the rate of utilization of the terminal acceptor NADP. Allen and Bennett (1981) have shown that the kinase activity is maximal during the induction phase of photosynthesis but decreases as CO_2 fixation proceeds at the higher steady-state level. Furthermore, Markwell et al. (1982) have shown that the kinase is inhibited by ADP. Horton and Foyer (1983) have examined the phosphorylation of the light-harvesting chlorophyll protein by the thylakoid protein kinase in the reconstituted chloroplast system (Chapter 4). The level of phosphorylation by ^{32}P ortho-phosphate was found to be maximal at high light intensity and in the absence of 3-phosphoglyceric acid. Dephosphorylation resulted from a decrease in light intensity or from the addition of 3-phosphoglyceric acid. Addition of ribose-5-phosphate, which acts as an ATP sink, also caused dephosphory-lation. It was concluded that the degree of phosphorylation is dependent on the redox and energy states of the system, thereby providing a mechanism for adapting light harvesting to the demands of carbon assimilation.

State 1–state 2 transitions do not cause drastic conformational changes in the thylakoid. State 1–state 2 changes have been reported to cause a decrease in the amount of stacked membranes in *Chlamydomonas* and *Egeria densa*. The mechanism by which phosphorylation produces state 1–state 2, transition has been proposed by Barber (1982) to involve alterations in the electrical properties at the periphery of the grana partition gaps. Phosphorylation is considered to change the overall balance of electrical charge on the exposed segment of the LHC–PS2 complex in the membrane. The phosphorylation has been calculated to increase the net negative charge on the exposed surface of the LHC–PS2 complex by about one electronic charge per 1.5 nm^2. This is large enough to cause coulombic repulsion between the phosphorylated surfaces at the periphery of the partition gap and bring about partial unstacking. This would allow migration of the phosphorylated LHC protein into the nonappressed membranes where, by random diffusion and collision, energy transfer to PSI is favored. Thus phosphorylation could act as a means of regulating the spatial relationships between functionally active proteins (Figure 2.7).

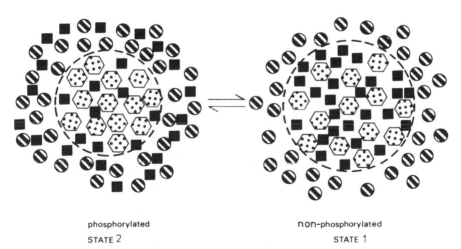

phosphorylated   non-phosphorylated

STATE 2   STATE 1

Figure 2.7. Diagrammatic representation of a surface view of a thylakoid membrane showing the regulation of state transitions by the action of the plastoquinol activated protein kinase. Dotted line denotes areas of membrane appression; hexagons, LHC–PSII complexes; circles, PSI complexes; squares, free LHC-II. (Taken from Horton, P. (1983). *FEBS Lett.* **152**, 47–52).

2.9. ELECTRON TRANSPORT IN CHEMOAUTOTROPHIC BACTERIA

The purple sulfur bacteria and the purple nonsulfur bacteria contain either bacteriochlorophyll-a or bacteriochlorophyll-b associated with the photosynthetic membranes that arise from the cytoplasmic membrane. The green sulfur bacteria contain either bacteriochlorophyll-c, -d, or -e in addition to bacteriochlorophyll-a. Most of the bacterial photosynthetic pigments are located in one or more spectrally different carotenoid–bacteriochlorophyll protein complexes. These may be divided functionally into reaction center complexes that are photochemically active and the antennae bacteriochlorophyll complexes that function in light absorption and transfer the absorbed energy to the reaction center complexes. The reaction centers of photosynthetic bacteria are usually isolated as complexes consisting of three polypeptides. These are labelled L, M, and H subunits, based on their migrations on SDS PAGE, and which in *Rhodopseudomonas sphaeroides*, for example, have molecular weights of 21, 24, and 28 kD. The purple bacteria contain a reaction center complex that retains photochemical activity after purification. This protein complex has a molecular weight of $\sim 100,000$ and contains the three distinct polypeptides in a 1:1:1 ratio. In addition, the complex contains four molecules of bacteriochlorophyll, two molecules of bacteriophaeophytin, one carotenoid, one or two quinone molecules, and an iron atom (Feher and Okamura, 1977). Most bacterial reaction centers contain nonheme iron in stoichiometric quantities and it is generally accepted that it functions in the path of electron transfer between the primary and secondary electron acceptors. Two of the bacteriochlorophyll molecules are the reaction center "special pair," P_{870}. At least one of the other chlorophyll molecules and one phaeophytin molecule act as intermediate electron carriers between P_{870} and an iron–quinone complex in the primary charge separation (Sauer, 1979). The antenna pigments are localized in one or two types of carotenoid–bacteriochlorophyll proteins. These are a B_{870} complex, in which the bulk chlorophyll absorbs at 870 nm, and a $B_{800-850}$ complex. The B_{870} complex is considered to be present in most if not all purple bacteria. Like the $B_{800-850}$ complex it has a rather simple composition. The B_{870} complex is found to consist of a 19,000 polypeptide with which two bacteriochlorophylls and two carotenoid molecules are associated. The $B_{800-850}$ complex is only found in some species and contains two nonidentical polypeptides in a ratio of 1:1. One polypeptide is associated with two exciton-coupled bacteriochlorophyll molecules absorbing at 850 nm while the other polypeptide contains one bacteriochlorophyll molecule, absorbing at 800 nm, and also one carotenoid.

The green bacteria are comprised of two families, the chlorobiaceae and the chloroflexaceae, which have similar morphology and pigment composition. The thylakoid membrane system of the green bacteria contains bacteriochlorophyll-a and either bacteriochlorophyll-c, -d, or -e as an accessary pigment. Of the pigment–protein complexes identified and purified from green bacteria only a water-soluble bacteriochlorophyll-a–protein antenna complex representing approximately 5% of the antenna pigment system has been fully characterized. This complex is a trimer of 140,000 molecular weight consisting of 42,000 molecular weight subunits. These take the form of rather flattened hollow cylinders that are structured such that the solvent exposed surface is composed of 15 strands of a β-pleated configuration described as "a string bag for holding pigment molecules." The folded protein encases seven bacteriochlorophyll molecules, the porphyrin rings of which are spaced approximately 12 Å from their nearest neighbor. The phytyl tails of the bacteriochlorophyll lie close together forming a hydrophobic core in the center of the subunit (Matthews et al., 1979).

Cyclic electron flow in photosynthetic bacteria does not result in oxygen evolution and is photochemically a much more simple process than that which is involved when water is oxidized. Only one type of photochemical reaction is present and cyclic electron flow is the principal means of ATP generation. Electrons from oxidizable substrates are only required for NADH production and also occasionally to maintain the redox balance of the cyclic process. Light energy is very efficiently transferred through the bacteriochlorophyll antenna system to the reaction center chlorophylls, for example, P_{870} in purple bacteria and P_{840} in green sulfur bacteria.

When subjected to brief flashes of light P_{870}^+ is generated within 3 psec and within 10 psec a radical pair is formed between P_{870} and bacteriopheophytin. From bacteriopheophytin the electron is transferred to an acceptor Q_A or Q_1, which is generally a tightly bound ubiquinone molecule and is essential for the stabilization of charge separation. The electron is passed to one of a pool of mobile ubiquinone molecules that migrates across the membrane and simultaneously transfers protons with electrons across the membrane to facilitate ATP formation. The electrons in the ubiquinone pool may be utilized in a side chain to reduce NAD, via ferredoxin, or may be cycled back to the reaction center (see figure 1.5). Electrons are introduced into the chain from oxidizable substrates via specific dehydrogenase enzymes and low potential c-type cytochromes. These cytochromes are rapidly oxidized by $P_{870}+$ (i.e., microseconds) but their re-reduction is slow (i.e., seconds). They, therefore, cannot participate in rapid cycling of electrons. High potential c-type cytochromes are rapidly re-reduced after oxidation and directly participate in the re-reduction of the oxidized reaction center chlorophylls.

Primary charge separation in the chlorobiaceae resembles that in PSI and consists of the transfer of an electron from the primary donor P_{840}, that is probably a dimeric bacteriochlorophyll-a, to a porphyrin that is possibly bacteriophaeophytin-c. The electron is subsequently transferred to a bacteriochlorophyll-a monomer and finally to two iron sulfur centers acting in series. $P_{840}+$ oxidizes cytochrome-c in 90 microseconds. Electron transport in the chloroflexaceae which contain bacteriophaeophytin-a may be different to that in the chlorobiaceae and more closely resemble that occurring in the purple bacteria. *Chloroflexus aurantiacus* which is a thermophilic green bacterium considered to be a most primative photosynthetic organism has characteristics more similar to the purple rather than the green bacteria reaction centers.

REFERENCES

Allen, J. F. and Bennett, J. (1981), *FEBS Lett.* **128**, 67–71.

Allen, J. F., Bennett, J., Steinback, K. E., and Arntzen, C. J. (1981). *Nature* **291**, 1–5.

Amesz, J. (1977). In *Encyclopedia of Plant Physiology* (A. Trebst and M. Avron, eds.), Vol. 8, pp. 238–245. Springer-Verlag, Berlin.

Anderson, J. M. and Andersson, B. (1981). In *Photosynthesis* (G. Akoyunoglou, ed.), Vol. 111, Structure and Molecular Organization of the Photosynthetic Apparatus, pp. 23–31. Balaban International Science Services, Philadelphia.

Barber, J. (1982). *Ann. Rev. Plant Physiol.* **33**, 261–295.

Bendall, D. S. (1977). *Int. Rev. Biochem.* **13**, 41–78.

Bennett, J. (1979a). *Trend. Biochem. Sci.* **4**, 268–271.

Bennett, J. (1979b). *FEBS Lett.* **103**, 342–344.

Bennett, J. (1980). *Eur. J. Biochem.* **104**, 85–89.

Bonaventura, C. and Myers, J. (1969). *Biochim. Biophys. Acta* **189**, 366–383.

Bouges-Bocquet, B. (1980). *Biochim. Biophys. Acta* **594**, 85–103.

Branton, D. and Park, R. B. (1967). *J. Ultrastruct. Res.* **19**, 283–303.

Douce, R. and Joyard, J. (1979). *Adv. Bot. Res.* **7**, 1–117.

Emerson, R. (1958). *Ann. Rev. Plant Physiol.* **9**, 1–24.

Emerson, R. and Lewis, C. M. (1943). *Am. J. Bot.* **30**, 165–178.

Evans, M. C. W. and Cammack, R. (1975). *Biochem. Biophys. Res. Commun.* **63**, 187–193.

Feher, G. and Okamura, M. Y. (1977). In *The Photosynthetic Bacteria* (R. K. Clayton and W. R. Sistrom, eds.), pp. 349–386. Plenum Press, New York.

Gantt, E., Lipschultz, C. A., and Zilinskas, B. A. (1976). In *Brookhaven Symposia in Biology* (J. M. Olson and G. Hind, eds.), No. 28, pp. 347–357. Brookhaven National Laboratory, New York.

Govindjee and Wydrzynski, T. (1981). In *Photosynthesis II* (G. Akoyunglou, ed.), Electron Transport and Photophosphorylation, pp. 293–305. Balaban International Science Services, Philadelphia.

Hall, D. O. and Rao, K. K. (1977). In *Encyclopedia of Plant Physiology* (A. Trebst and M. Avron, eds.), Vol. 5, pp. 206–216. Springer-Verlag, Berlin.

Haupt, W. (1982). *Ann. Rev. Plant Physiol.* **33**, 205–233.

Henry, L. E. A. and Moller, B. L. (1981). *Carlsberg Res. Commun.* **46**, 227–242.

Hill, R. and Bendall, D. S. (1960). *Nature* **186**, 136–137.

Hill, R. and Rich, P. R. (1983). *Proc. Natl. Acad. Sci. USA* **80**, 978–982.

Horton, P. (1982). *Biochem. Soc. Trans.* **10**, 338–340.

Horton, P., Allen, J. F., Black, M., and Bennett, J. (1981). *FEBS Lett.* **125**, 193–196.

Horton, P. and Foyer, C. H. (1983). *Biochem. J.* **210**, 517–521.

Izawa, S., Heath, R. L., and Hind, G. (1969). *Biochim. Biophys. Acta* **80**, 388–398.

Joliot, P. and Kok, B. (1975). In *Bioenergetics of Photosynthesis* (Govindjee, ed.), pp. 387–412. Academic Press, New York.

Katoh, S. (1977). In *Encyclopedia of Plant Physiology* (A. Trebst and M. Avron, eds.), Vol. 5, pp. 247–252. Springer-Verlag, Berlin.

Ke, B., Dolan, E., Suhagara, K., Hawkridge, F. M., Demeter, S., and Shaw, E. R. (1977). In *Photosynthetic Organelles* (S. Miyachi, , S. Katoh, Y. Fujita, and K. Shibata, eds.), Special Issue *Plant Cell Physiol.* **3**, 187–199.

Kirk, J. T. O. and Tilney-Basset, R. A. E. (1978). In *The Plastids. Their Chemistry, Structure, Growth and Inheritance*, 2nd ed. Elsevier/North Holland Biomedical Press, Amsterdam.

Leech, R. M., Thomson, W. W., and Platt-Aloia, K. A. (1981). *New Phytol.* **87**, 1–9.

Malkin, R. (1982). *Ann. Rev. Plant Physiol.* **33**, 455–479.

Markwell, J. P., Baker, N. R., and Thornber, J. P. (1982). *FEBS Lett.* **142**, 171–174.

Murata, N. (1969). *Biochim. Biophys. Acta* **189**, 171–181.

Nugent, J. H. A. and Evans, M. C. W. (1979). *FEBS Lett.* **101**, 101–104.

Nugent, J. H. A., Møller, B. L., and Evans, M. C. W. (1981). *Biochim. Biophys. Acta* **634**, 249–255.

Possingham, J. V. (1980). *Ann. Rev. Plant Physiol.* **31**, 113–129.

Radmer, R. and Cheniae, G. (1977). In *Topics in Photosynthesis* (J. Barber, ed.), Vol. 2, Primary Processes of Photosynthesis, pp. 303–348. Elsevier, Amsterdam.

Renger, G. (1978). In *Photosynthetic Water Oxidation* (H. Metzner, ed.), pp. 229–248. Academic Press, London.

Rich, P. R. and Bendall, D. S. (1981). In *Photosynthesis* (G. Akoyunoglou, ed.), Vol. II, Electron Transport and Photophosphorylation, pp. 551–558. Balaban International Science Services, Philadelphia.

Sauer, K. (1979). *Ann. Rev. Phys. Chem.* **30**, 155–178.

Staehelin, L. A. (1981). In *Photosynthesis* (G. Akoyunoglou, ed.), Vol. III, Structure and Molecular Organization of the Photosynthetic Membrane, pp. 3–14. Balaban International Science Services, Philadelphia.

Staehelin, L. A., Armond, P. A., and Miller, K. R. (1976). In *Brookhaven Symposia in Biology* (J. M. Olson and G. Hind, eds.), No. 28, pp. 278–315. Brookhaven National Laboratory, New York.

Steinback, K. E., Burke, J. J., and Arntzen, C. J. (1979). *Arch. Biochem. Biophys.* **195**, 546–557.

Stemler, A. (1978). In *Photosynthetic Oxygen Evolution* (H. Metzner, ed.) pp. 283–293. Academic Press, London.

Thornber, J. P. and Markwell, J. P. (1981). *Trend. Biochem. Sci.* **6**, 122–125.

Wildman, S. G., Hongladarom, T., Honda, S. I. (1966). *Organelles in Living Plant Cells*, 16 mm sound film. Educational Film Sales and Rentals, University of California, Berkeley.

3

INTERACTIONS BETWEEN THE THYLAKOIDS AND STROMA

3.1. ELECTRON TRANSPORT

Photophosphorylation and reduction of nicotinamide nucleotides, ferredoxin, and so on take place at the boundary between the photosynthetic membranes and the stroma and require the interaction of both of these.

In the chemoautotrophic bacteria the simplicity of the cyclic photosynthetic electron transport chain system encompassing a single photosystem is generally offset by the physical proximity and association of the respiratory electron transport chain. In higher plants and algae the photosynthetic and respiratory electron transport chains are spatially separated but these organisms possess two types of photosystem and electron flow can be considerably more complex. When water is the primary donor for electron transport and NADP is the ultimate electron acceptor the overall process is one of "noncyclic" electron flow (Figure 3.1), which can be represented as

$$2H_2O + 2NADP \xrightarrow{\quad 8h\nu \quad} O_2 + 2NADPH + 2H^+ \qquad (3.1)$$

and ATP synthesis coupled to this system is referred to as noncyclic photophosphorylation. NADP is reduced directly by PSI, via ferredoxin and ferredoxin–NADP reductase. In green sulfur bacteria the reduction of NAD ($NAD^+/NADH$ $E_m = -0.35$ V) is brought about by a similar noncyclic pathway using a soluble ferredoxin and a ferredoxin–NAD reductase. Electrons derived from oxidizable substrates such as hydrogen sulfide and thiosulfate, enter the pathway via specific reductases at cytochrome 551 ($S_2O_3^{2-}$) and cytochrome 553 (S^{2-}) and pass through cytochrome 555 to the reaction center chlorophyll P_{840}. A noncyclic pathway generates NADH while the cyclic pathway generates ATP. Purple photosynthetic bacteria employ a rather different means of generating sufficient energy to reduce NAD. The high energy state of the membrane ($\sim$) provides not only the conditions to drive ATP synthesis but also the energy for NAD reduction. Electrons from oxidizable substrates are transferred to the ubiquinone pool via specific dehydrogenases. The high energy state of the membrane makes it possible for electrons from the ubiquinone pool to reduce NAD through energy-linked reverse electron flow.

In oxygenic photosynthesis electrons are channeled through the electrochemical gradient via the action of the two light reactions to reduce NADP. The electron transport chain has a high affinity for NADP, which is a highly

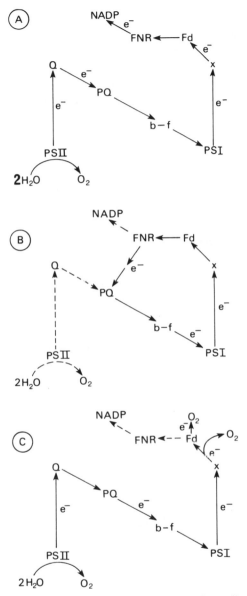

Figure 3.1. Diagram to show the paths of (A) noncyclic, (B) cyclic, and (C) pseudocyclic electron flow in chloroplasts.

efficient electron sink. This noncyclic process generates the ATP and NADPH that are used to drive the biosynthetic reactions of the chloroplast and cytoplasm and in particular the fixation of carbon dioxide in the RPP pathway in the stroma. If the H^+:e ratio is 2 then 4 quanta absorbed per photosystem will transport eight protons into the intrathylakoid space while generating two NADPH in the stroma. The coupling factor ($CF_0 - CF_1$) converts one ADP to one ATP for every three protons transported and therefore eight protons are theoretically not quite sufficient to generate three ATP per two NADPH. The actual number of ATP synthesized per pair of electrons transported through the photosynthetic electron transport chain (the ATP:$2e$ ratio) is still a matter of debate. Numerous investigations have shown that the ATP:$2e$ ratio during noncyclic electron flow is probably variable, ranging from 1.0 (Arnon and Chain, 1977) to 2.0 (Robinson and Wiskich, 1976).

The fixation of one molecule of carbon dioxide into a hexose sugar via the RPP pathway requires three molecules of ATP and two molecules of NADPH (Chapter 4). Carbon fixation alone requires that the ATP:$2e^-$ ratio is at least 1.5 for noncyclic photophosphorylation. If this is the case then 8 light quanta, by removing four electrons from water to NADPH, should be sufficient to synthesize three ATP molecules. In fact, measurements have shown that the quantum requirement for CO_2 reduction is usually higher than the 8 light quanta necessary for noncyclic electron flow. It is therefore most likely that the ATP requirements of the chloroplast cannot be met by noncyclic electron flow alone. In addition to the RPP pathway and the energy requirements of other biosynthetic pathways in the chloroplast (Chapter 6), ATP and reducing power are rapidly transferred via indirect shuttle systems to the cytoplasm. Thus situations must arise when it is required that more ATP be generated relative to NADPH. There are two pathways of electron flow capable of supplying extra ATP without NADPH. These are the "cyclic" and "pseudocyclic" pathways (the latter involves reduction of molecular oxygen). The chloroplast attempts to optimize its ATP and NADPH production in accord with the metabolic needs of the stroma. It does so by switching between noncyclic and cyclic pathways of electron flow. The concurrent operation of cyclic and noncyclic photophosphorylation could yield three ATP and two NADPH with an optimal total requirement of 12 light quanta. It is most probable that the feedback control mechanism that acts to produce this change in electron flow is the redox poise of the NADPH:NADP pool.

3.2. CYCLIC ELECTRON FLOW AROUND PSI

Cyclic electron flow around PSI is a process in which electrons from reduced ferredoxin or NADPH are transferred back to the plastoquinone pool and are then returned to P_{700} via the cytochrome b–f complex and plastocyanin (Figure 3.1) thus forming a redox loop around PSI. The PMF that drives cyclic photophosphorylation is generated by proton shuttling across the membrane resulting from the vectorial oxidation and reduction of plastoquinone. The situation may be more complex than this because of the manner of interaction between the plastoquinone pool and the cytochrome b–f complex. This interaction may take the form of a type of Q cycle as described by Mitchell (1976) in which electron transfer across the membrane must be followed by an electroneutral return of reducing equivalent after uptake of H^+ by the plastoquinone pool at the stromal surface (Crowther and Hind, 1981). Such a mechanism that is unlikely to operate in linear electron flow when an efficient electron acceptor is available has been suggested to account for the slow rise in the electric field indicating bandshift at 518 nm (termed $P_{518 \text{ slow}}$), which is observed following a single turnover flash of light.

The transthylakoid pH gradient appears to limit the overall rate of electron flow while the controlling factor in the transition between linear and cyclic electron flow seems to be the specific reduction of a component of the thylakoid membrane. Feedback control of electron flow, by redox poise (NADPH:NADP) is considered to be particularly important in the control of the cyclic pathway. When NADP is available in its oxidized form it traps electrons and thereby inhibits cyclic electron transport. In noncyclic electron flow if less ATP is generated during NADP reduction than is required for the conversion of CO_2 into sugars in the RPP pathway, then ATP will limit NADPH oxidation. Similarly, the magnitude of the transmembrane proton gradient effects the rate of reoxidation of the plastoquinone pool and thus may regulate the overall rate of electron flow. In these situations NADPH may be expected to accumulate and the NADP pool depleted. NADP will therefore not be available to accept electrons that will then be diverted into the cyclic and pseudocyclic pathways. High NADPH:NADP ratios partially close PSII reaction center traps because of the reduction of the plastoquinone pool. This is indicated by a bandshift at 550 nm (called $C550$) that reflects the redox state of the electron acceptor Q. Arnon and Chain (1977) proposed that electron flow from NADPH to C550 may function as a physiological redox poising mechanism that occurs when NADPH accumulates. The partial closure of the PSII traps caused by plastoquinone reduction would attentuate electron flow from PSII. The rate of cyclic electron flow can be considered to be a function of the electron "pressure" between the photosystems, the excitation of PSI and the oxidation state of the electron carriers. The relative activities of the noncyclic and cyclic pathways would therefore also be determined to some extent by the redox state of the plastoquinone pool. The operation of the cyclic pathway may therefore be considered to be subject to a fine redox control that may involve either an attenuated electron flow from PSII or the oxidation of electron carriers such as ferredoxin by molecular oxygen, which prevents overreduction of the electron carriers of the cyclic system. Inhibition of the pathway is caused by oxidation or by overreduction of the electron carriers. For example, in the absence of oxygen, cyclic electron flow is inhibited by illumination with red light, which excites PSII, because of overreduction. This effect can be alleviated by the addition of oxygen, which drains excess electrons from the pathway. When PSI alone is excited by far-red light, oxygen causes inhibition by the oxidation of the electron carriers. Cyclic electron flow must therefore be linked to an electron donor system such as PSII to be functional in the presence of oxygen.

Cyclic electron flow appears to be an important means of generating ATP in the "induction" or "lag" phase of photosynthesis (Chapter 4). In the lag phase the demand for ATP to phosphorylate sugar intermediates is high and an increased ATP:NADPH ratio may be required. The slow rise in the flash induced electrochromic shift at P_{518} is seen generally only in association with cyclic electron flow and has therefore been used to monitor this process. During the lag phase the pronounced slow rise in P_{518} is greater than is seen in steady-state photosynthesis indicating increased cyclic electron flow during this period. Cyclic photophosphorylation is sensitive to inhibition by antimycin A. The presence of antimycin A during the lag phase of photosynthesis decreases the rate of ATP synthesis and lengthens the induction period (Chapter 4). In steady-state photosynthesis cyclic electron flow participates in ATP synthesis mainly at moderate or high light intensities. When the uncoupler, NH_4Cl, is added during steady-state carbon assimilation the slow rise in the P_{518} electrochromic shift is increased as is cytochrome turnover. This suggests that cyclic electron flow is increased by the presence of the uncoupler, to pump additional protons, in order to maintain ΔpH against the uncoupling action of ammonium chloride.

The leaves of plants that exhibit the C4-type of photosynthesis require five ATP per two NADPH in the assimilation of carbon dioxide to produce sugar phosphate and therefore should always require ATP additional to that provided by noncyclic electron flow. Even the very best experimentally observed ATP:$2e$ ratios of 2 would not be sufficient to meet the requirements of CO_2 fixation in these plants. However, this overall energetic requirement is compartmented. C4 photosynthesis involves the condensation of CO_2 with phosphoenolpyruvate in the mesophyll cells to form oxaloacetate. Oxaloacetate is reduced to malate, in the mesophyll cells, which is then transported to the bundle sheath cells where it is decarboxylated to yield CO_2, NADPH, and pyruvate. Pyruvate is then transported back to the mesophyll cells where the cycle is completed by its conversion to phosphoenolpyruvate. CO_2 is fixed in the bundle sheath cells by ribulose-1,5-bisphosphate (RuBP) carboxylase as in C3 photosynthesis (Chapter 4). When the energetic requirements of this pathway are considered it is apparent that the conversion of one molecule of CO_2 into triosephosphate requires the generation of two ATP in the bundle sheath chloroplast and three ATP and two NADPH in the mesophyll chloroplast (Chapter 7). The mesophyll chloroplasts have the same overall energetic requirement as chloroplasts from a C3 plant such as spinach.

The bundle sheath chloroplasts of maize are deficient in PSII. During CO_2 fixation in bundle sheath cells NADPH formation as a result of noncyclic electron flow is very limited and unlikely to be very important in PGA reduction. Cyclic electron flow around PSI has been shown to be the major electron transfer pathway in bundle sheath chloroplasts. This generates the additional ATP required for C4 photosynthesis. The PSII present in bundle sheath chloroplasts is insufficient to poise cyclic electron flow and it has been suggested that NADPH, generated by malate decarboxylation, provides the necessary reducing power (Chapman et al., 1980; Leegood et al., 1983). In the mesophyll chloroplasts cyclic electron flow can be poised by the donation of electrons by PSII and oxidation by molecular oxygen.

3.3. CYTOCHROME *b*559 AND CYCLIC ELECTRON FLOW AROUND PSII

The thylakoid membranes of eukaryotic cells contain two *b*-type cytochromes with α absorption bands at 559 nm. These two forms may be distinguished by their midpoint potentials. High potential cytochrome $b559$ ($b559_{HP}$) has a midpoint potential of $+330$ to $+400$ mV at pH 7–8 and accounts for approximately 60% of the total cytochrome $b559$. Low potential cytochrome $b559$ ($b559_{LP}$) has a midpoint potential of $+50$ to $+100$ mV at pH 7–8. Cytochrome $b559_{HP}$ is closely associated with PSII, whereas cytochrome $b559_{LP}$ is found in association with cytochromes $b563$ and f in isolated PSI particles. In freshly prepared chloroplasts there are approximately two equivalents of cytochrome $b559_{HP}$ per reaction center. The very positive potential of $b559_{HP}$ has been suggested to result from the hydrophobic environment close to positive charges in which the cytochrome resides. When O_2 evolution is impaired cytochrome $b559_{HP}$ can be oxidized by PSII, and a correlation has been demonstrated between the manganese content and the level of this cytochrome in chloroplasts. Mild heating, aging, and detergents cause the conversion of $b559_{HP}$ to a lower potential form. The role of cytochrome $b559_{HP}$ remains unclear. It has been suggested to function in a safety mechanism that would keep P_{680} in the reduced form in the event of loss of activity of the water-splitting complex (Bendall, 1982). Several workers have concluded from the redox behavior of $b559_{HP}$ that this cytochrome is directly involved in cyclic electron flow around PSII. The addition of compounds such as tetraphenylboron or carbonyl cyanide-p-trifluoromethoxyphenylhydrazone (FCCP) is necessary in order to demonstrate turnover of cytochrome $b559_{HP}$ in chloroplasts (Heber et al., 1979). The oxidation of cytochrome $b559_{HP}$ when water oxidation is inhibited suggests that $b559_{HP}$ competes in electron donation to PSII. However, such treatments also may result in conversion to the low potential form. A small proportion of the total cytochrome $b559$ (10%) is apparently phosphorylated by a protein kinase in the light.

3.4. PSEUDOCYCLIC ELECTRON FLOW

Since the first experiments of Hill (1939) it has been repeatedly demonstrated that isolated thylakoids are able to reduce a variety of natural and artifical oxidants. Mehler (1951) observed that isolated illuminated thylakoids will slowly reduce molecular oxygen in the absence of added electron acceptors. This reaction, in which oxygen is both evolved and taken up by the action of the electron transport chain, is called the *Mehler* reaction or pseudocyclic electron flow (Figure 3.1). Oxygen is reduced by the electron transport chain during CO_2 fixation in plants and algae. NADP is reduced with a high quantum efficiency and is undoubtedly the preferred electron acceptor for photosynthesis. Oxygen is not an efficient electron acceptor but it could be a necessary one since electrons can be diverted to oxygen (for example, to poise cyclic electron flow) at times when NADP levels are low. Light-scattering measurements performed with intact leaves have shown that cyclic electron flow becomes possible only when electron flow to oxygen has been largely saturated. Oxygen reduction is saturated at low light intensities. As light intensity is increased the reduction of oxygen does not increase proportionally to CO_2-dependent oxygen evolution. At low light intensities CO_2 reduction is slow and electron transport to oxygen could be sufficient to supply the additional ATP required for CO_2 reduction. When NADP is largely reduced and electron flow to oxygen is near saturation, electrons are diverted into the cyclic pathway (Egneus et al., 1975). Even during cyclic photophosphorylation electron drainage to oxygen may be required as a regulatory device that is necessary to prevent overreduction of the electron carriers. Electron transport to oxygen poises the electron carriers of the cyclic pathway while being in itself a coupled process leading to ATP synthesis. Direct photoreduction of oxygen by the electron transport chain is inescapable in all organisms carrying out oxygenic photosynthesis. Since water is the donor for electron transport, the thylakoid membranes exist in an environment rich in oxygen. This oxygen competes with NADP for reducing equivalents (Figure 3.1) so that when the NADP is largely reduced electron transport to oxygen is stimulated. Electron flow to oxygen is generally very low and at most can only account for 16% of the net steady-state rate of oxygen evolution during CO_2 reduction in intact spinach chloroplasts at high light intensities. However, significant rates of oxygen reduction can be observed particularly when CO_2 fixation is limited, for example, during the initial lag phase of photosynthesis. Marsho and coworkers (1979) observed that when dark-adapted chloroplasts or cells, supplemented with saturating amounts of bicarbonate, were illuminated, oxygen evolution began immediately. However, this initial rate of oxygen evolution was counterbalanced by a simultaneous increase in the rate of O_2 uptake so that little net O_2 was evolved or consumed during the first minute of illumination. After this induction phase the rate of oxygen uptake diminished to a low level while the rate of oxygen evolution increased. Observed rates of oxygen reduction during steady-state photosynthesis in algae appear to be somewhat greater than those observed in higher plant chloroplasts. In intact cells of the cyanobacterium *Anacystis nidulans* the onset of illumination has been found to result in a transient burst of H_2O_2 production, which after about 5 min slows to a continuous steady rate of H_2O_2 excretion (Patterson and Myers, 1973). Hydrogen peroxide is sometimes found to be excreted from isolated intact chloroplasts during photosynthesis (Steiger and Beck, 1981).

The structure of the oxygen molecule requires that reduction is preferentially a univalent process. The primary side of oxygen reduction is likely to be on the reducing site of PSI where the electrochemical potential is favorable. The interaction of oxygen with a light generated PSII reductant has been observed in chloroplasts and algae although this reaction usually occurs at a very low rate. The reducing side of PSI is able to donate single electrons to oxygen in an autoxidative process. The PSI electron transport chain contains at least two autoxidizable species (Lien and San Pietro, 1979). The most likely sites of oxygen reduction are reduced ferredoxin (Allen, 1977a) and the iron–sulfur centers (Asada and Nakano, 1978). The addition of a single electron to oxygen produces the superoxide radical anion O_2^-. This radical is regarded as the primary agent of oxygen toxicity because superoxide can give rise to highly reactive oxygen species (e.g., OH) particularly in the presence of metal ions (Elstner, 1982). In an aqueous environment, superoxide is not a particularly reactive molecule as it becomes strongly solvated by water molecules and is therefore a poor nucleophile. In aqueous solution O_2^- can act either as a reducing agent, for example, it can reduce cytochrome f and plastocyanin, or it can act as a weak oxidizing agent, for example, it can slowly oxidize thiol-containing compounds. In the absence of a suitable oxidant two superoxide molecules will spontaneously dismutate in the following reaction:

$$O_2^- + O_2^- + 2H^+ \longrightarrow H_2O_2 + O_2 \qquad (3.2)$$

The reaction is slow at neutral or alkaline pH values, being most rapid at acidic pH values when O_2^- is protonated. The dismutation of superoxide is greatly accelerated by the presence of a Cu–Zn containing superoxide dismutase in the chloroplast stroma.

3.5.   SUPEROXIDE DISMUTASE

The enzyme superoxide dismutase is present in all aerobic cells and plays an essential role in allowing organisms to survive in the presence of oxygen

(Fridovich, 1978). Superoxide dismutases are metalloproteins containing iron, or manganese, or copper and zinc as the prosthetic group. The Cu–Zn enzyme that is found in all eukaryotic cells contains two atoms of copper and two of zinc per dimer and is inhibited by cyanide and H_2O_2. The manganese superoxide dismutases that are found in bacteria and in the mitochondrial matrix of plants and animals can contain from one to four manganese atoms and are insensitive to cyanide and H_2O_2. The iron superoxide dismutases predominate in prokaryotic organisms but have been reported in three families of higher plants: the Ginkgoaceae, Cruciferae, and Nymphaceae (Salin and Bridges, 1981, 1982). These Fe-superoxide dismutases have either one or two iron atoms per molecule and are cyanide insensitive but inhibited by H_2O_2. The Cu–Zn enzymes are absent in prokaryotic and most eukaryotic algae that contain either manganese and/or iron enzymes. Higher plants generally contain large amounts of the Cu–Zn enzyme and a smaller amount of the manganese enzyme.

Chloroplasts isolated from the leaves of higher plants contain large quantities of superoxide dismutase activity (Jackson et al., 1978). This, in itself, is an indication that the chloroplast is often faced with the problem of superoxide production. Most of the chloroplast dismutase is of the cyanide-sensitive Cu–Zn type. The bulk of this enzyme is freely soluble in the stroma although some can associate with the thylakoid membrane (Van Ginkel and Brown, 1978). Although a cyanide-insensitive manganese enzyme is often found associated with isolated intact chloroplasts it is located outside these organelles and may be due to mitochondrial contamination. Leaf mitochondria that contain approximately 5% of the total superoxide dismutase activity have the manganese enzyme in the matrix. In addition to the presence of superoxide dismutase in the chloroplasts, the thylakoid membranes contain manganese at a concentration of approximately 1-g atom of manganese per $50-100$ chlorophylls. Free Mn^{2+} is an inhibitor of lipid peroxidation in membranes and will inhibit superoxide-mediated reactions in a manner similar to superoxide dismutase. The LHC protein has been shown to contain bound manganese. There at at least two binding sites for manganese in the LHC that have different affinities for Mn^{2+}. There is a small pool of high affinity binding sites (equivalent to 1 Mn per 26,000 polypeptide) and a much larger pool of low affinity binding sites (Figure 3.2). The binding to the low affinity sites is inhibited by the presence of sulphydryl binding reagents (e.g., N-ethyl maleimide) and is competitively inhibited by Mg^{2+}. The presence of bound manganese on the purified LHC confers a cyanide-in-

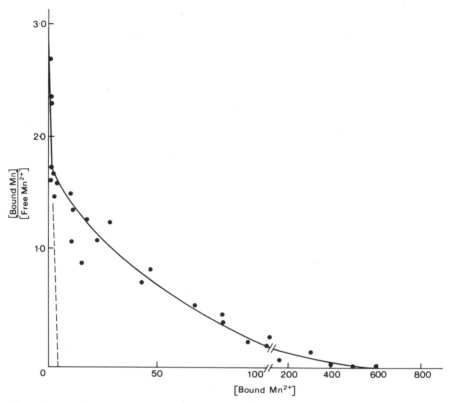

Figure 3.2.   Scatchard plot of the binding of Mn^{2+} to purified LHC binding protein from spinach. Binding of Mn^{2+} was measured at room temperature using electron spin resonance. Reaction mixtures contain 10 μg chlorophyll/mL (chlorophyll-a/b = 1.2) and the appropriate concentration of $MnCl_2$ in a total volume of 1 mL.

sensitive superoxide dismutase-like activity on the protein. Since the LHC readily binds cations such as Mg^{2+}, Ca^{2+}, and Mn^{2+}, it is probable that it will bind manganese *in situ* and may provide a means for the removal of superoxide in the membrane.

3.6.   THE REMOVAL OF H_2O_2

The product of the superoxide dismutation reaction [equation (3.2)] is hydrogen peroxide, H_2O_2, an intermediate that the chloroplast appears to consider less hazardous than O_2^-. However, H_2O_2 is a potent inhibitor of photosynthesis; CO_2 fixation is inhibited by 50% in the presence of 10-μM H_2O_2 (Kaiser, 1976). The sites of this inhibition are primarily the fructose and sedoheptulose bisphosphatases of the Calvin cycle. These enzymes contain essential thiol groups that must be reduced for activity. H_2O_2 rapidly oxidizes these enzymes so that they are virtually inactive under physiological conditions (Heldt et al., 1977; Charles and Halliwell, 1980). In addition, H_2O_2 slowly inactivates the chloroplasts Cu–Zn superoxide dismutase thereby reducing the ability of the chloroplast to remove O_2^-. A major toxic action of H_2O_2 lies in the ability of this compound to participate in chain reactions and yield oxygen derivatives of greater reactivity. The presence of O_2^- and H_2O_2 in biological systems leads to the formation of the most reactive radical known to organic chemistry, the hydroxyl free radical $\cdot$OH. This reactive molecule will destroy almost all biological molecules. In particular it can initiate lipid peroxidation when generated in the proximity of a membrane. The fatty acids of the chloroplast lipids are polyunsaturated and therefore susceptible to hydrogen abstraction. The formation of lipid peroxides causes fragmentation of the membrane and leakage of solutes. Lipid peroxides themselves are powerful inhibitors of enzymes and their disintegration to yield aldehydes and other products causes severe damage to membrane integrity. The chain reaction of lipid peroxidation is effectively inhibited by the abstraction of hydrogen from either a carotenoid pigment or α-tocopherol.

The reaction of H_2O_2 with O_2^- is catalyzed by iron salts, which are essential for the reaction of the two molecules to occur. The hydroxyl radical is produced by the following reactions:

$$Fe^{3+} + O_2^- \longrightarrow Fe^{2+} + O_2 \qquad (3.3)$$

$$Fe^{2+} + H_2O_2 \longrightarrow \cdot OH + OH^- \qquad (3.4)$$

$$\text{net } H_2O_2 + O_2^- \xrightarrow{\text{Fe salt}} O_2 + \cdot OH + OH^- \qquad (3.5)$$

Because of these properties of H_2O_2 it is essential that the illuminated chloroplast contains mechanisms for the rapid removal of this compound. Aerobic cells have evolved enzymes of two types, the catalases and the peroxidases to prevent the accumulation of H_2O_2. In leaf cells most of the catalase is localized in the matrix of the peroxisomes that are discrete organelles bounded by a single membrane. The peroxisomes contain H_2O_2-generating enzymes, which, when oxidizing substrates, produce considerable amounts of H_2O_2 that can escape catalase action and diffuse into the surrounding medium. Peroxisomes are often found in close physical contact with chloroplasts. It has been suggested that H_2O_2 generated in the chloroplasts could diffuse into the peroxisomes for destruction by catalase. However, catalase itself is not effective in dealing with low concentrations of H_2O_2 as the catalatic breakdown of H_2O_2 requires the formation of a catalase–H_2O_2 complex, which then reacts with a second molecule of H_2O_2. Thus two molecules of H_2O_2 must impinge on the active site of a single catalase molecule in the reaction:

$$2H_2O_2 \longrightarrow 2H_2O + O_2 \qquad (3.6)$$

In the presence of suitable substrates catalase also exhibits peroxidatic activity as follows:

$$RH_2 + H_2O_2 \longrightarrow R + 2H_2O \qquad (3.7)$$

Where RH_2 is an oxidizable substrate. In general catalases are tetrametic haemoproteins with a high catalatic and a relatively low peroxidatic activity.

Peroxidase enzymes utilize H_2O_2 to oxidize a wide variety of substrates in a reaction of the type shown in equation (3.7). In animal tissues H_2O_2 can be removed by both catalase and a specific peroxidase that reduces glutathione but these enzymes are not present in chloroplasts. Catalase is often found associated with isolated chloroplasts (Allen, 1977b), but this activity is largely due to contamination by peroxisomes in crude fractions. There are also numerous reports of catalase activity in chloroplasts, mitochondria, and cytoplasm, but it has not been unambiguously demonstrated that catalase detected in these fractions after subcellular fractionation is not derived from damaged peroxisomes or due to nonspecific catalatic activities such as that of heme proteins. Plant tissues do not contain the selenium metalloprotein, glutathione peroxidase, that occurs in animal cells. They do display a wide range of peroxidatic activity, which has a broad specificity for substrates. These peroxidases are generally assayed using artifical substrates such as O-dianisidine and the natural substrates for these enzymes *in vivo* are often not known. Isolated chloroplasts contain very little peroxidase activity of this type but do have an ascorbate specific peroxidase that is membrane bound (Groden and Beck, 1979). High levels of a soluble ascorbate peroxidase have been observed in plants (Kelly and Latzko, 1979).

L-ascorbic acid, a hexose sugar derivative, is ubiquitous in the plant kingdom. It is essential to plant metabolism and inhibition of its synthesis prevents growth. In the chloroplast stroma ascorbic acid is present at concentrations of up to 50 mM. This large pool of ascorbic acid appears to be directly involved in the detoxification and removal of deleterious oxygen derivatives. Ascorbic acid will react with superoxide in the reaction:

$$O_2^- + 2H^+ + \text{ascorbate} \longrightarrow \text{dehydroascorbate} + H_2O_2$$
$$(3.8)$$

This reaction can proceed at rates comparable with that of superoxide dismutase in tissues where the ascorbic acid concentration is high. Ascorbic acid is a quencher of singlet oxygen and will rapidly react with hydroxyl radicals. It can also react nonenzymically with H_2O_2 in crude chloroplast extracts. Ascorbate peroxidases that catalyze the reaction.

$$H_2O_2 + \text{ascorbate} \longrightarrow \text{dehydroascorbate} + 2H_2O \quad (3.9)$$

have been found in several higher plants and algae. The most characterized of these is the ascorbate peroxidase of *Euglena gracilis*, which is a haemoprotein like many other peroxidases. The enzyme display high affinities both for ascorbate ($K_m = 410 \ \mu M$) and H_2O_2 ($K_m = 56 \ \mu M$). The insoluble spinach chloroplast enzyme appears to have very similar characteristics. The amount of ascorbate peroxidase activity detectable in chloroplasts fluctuates as does the ascorbic acid content, changing with the seasons and with the age of the leaf. These facts could account for discrepancies in the results of experiments on H_2O_2 generation by illuminated chloroplasts. When ascorbate peroxidase and ascorbic acid concentrations are maximal, H_2O_2 could be effectively removed as it was formed during illumination of intact chloroplasts. Maximal rates of photosynthesis may then be achieved without the addition of catalase. H_2O_2 would accumulate during photosynthesis at times when the ascorbic acid content and ascorbate perioxidase of chloroplasts were low. In this situation the addition of catalase would stimulate the apparent rate of photosynthesis.

Ascorbic acid, in plant tissues, is nearly always found in the reduced state and very little is found to be present in the oxidized form, dehydroascorbic acid (Foyer, et al., 1983). During illumination of leaves, protoplasts, and isolated chloroplasts, the ascorbate content is maintained at approximately the same concentration as in the dark even in the absence of CO_2 when pseudocyclic electron flow would be expected to be maximal (Figure 3.3). Dehydroascorbate, if left in the oxidized state, will rapidly decompose to yield, among other products, oxalic acid. Photoxidation of ascorbate to

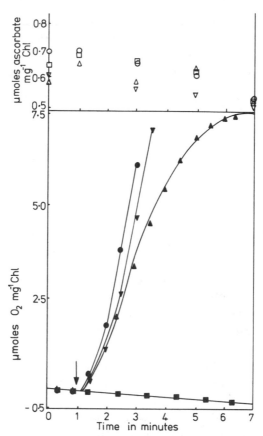

Figure 3.3. The effect of illumination on the ascorbate content of intact spinach chloroplasts in the presence of different bicarbonate concentrations. CO_2-dependent O_2 evolution and O_2 uptake were measured in chloroplasts illuminated in the oxygen electrode in the absence (□) and the presence of either 0.25-mM NaHCO$_3$ (△), 5.0-mM NaHCO (∇)$_3$, or 5.0-mM NaHCO$_3$, and 200-U catalase (○). Ascorbate content was measured at the same time in the absence (■) and the presence of either 0.25-mM NaHCO$_3$ (▲), 5.0-mM NaHCO$_3$ (▼), or 5.0-mM NaHCO$_3$, and 200-U catalase (●). Light on (↓). (Taken from Foyer, C., Rowell, J., and Walker, D. (1983). *Planta* **157**, 239–244).

oxalate by illuminated thylakoids has been observed. There, therefore, must be an efficient reducing mechanism(s) that actively maintains ascorbic acid in the reduced state, inspite of its rapid oxidation by oxidizing catalysts, ascorbate oxidases, O_2^-, and so on.

The chloroplast stroma contains millimolar concentrations of the simple tripeptide glutathione [L-Y-glutamyl-L-cysteinylglycine (GSH).] Glutathione is probably the most abundant natural low molecular weight thiol compound.

A principal function of GSH is to protect proteins and cell membranes against free radicals and peroxides and to maintain the sulphydryl groups of enzymes in their reduced state. It can be used to stabilize some of the enzymes of the Calvin cycle *in vitro*. GSH is a stronger reducing agent than ascorbic acid and can react with hydroxyl radicals and superoxide although in the latter case it is less effective than ascorbate.

Glutathione will react nonenzymically with dehydroascorbate at pH values above 7.0. The chloroplast stroma becomes alkaline upon illumination reaching approximately pH 8.0. Dehydroascorbate is rapidly reduced by GSH at pH 8.0 in the reaction:

$$2GSH + \text{dehydroascorbate} \longrightarrow \text{ascorbate} + GSSG$$
$$\text{(2 cysteine)} \qquad\qquad\qquad\qquad\qquad \text{(cystine)}$$
$$(3.10)$$

where GSSG is the oxidized form of glutathione. An enzyme that catalyzes the above reaction, dehydroascorbate reductase (alternatively called GSH dehydrogenase), EC 1.8.5.1, is present in many plant tissues. Dehydroascorbate reductase was not initially detected in spinach chloroplasts (Foyer and Halliwell, 1976), but has now been shown to be present in pea and spinach chloroplasts (Jablonski and Anderson, 1981). This enzyme may show seasonal variations as does ascorbate peroxidase. The reduction of dehydroascorbate is necessary under oxidizing conditions, especially in the presence of metal ions, because ascorbic acid itself can produce toxic derivatives. Autoxidizing ascorbate in the presence of traces of metal ions such as Cu^{2+} can produce highly reactive chemical species that can combine with and damage proteins, can cleave DNA and RNA, and can initiate lipid peroxidation. Yet in the reduced form ascorbic acid inhibits lipid peroxidation.

GSSG is converted back to the reduced form by the enzyme glutathione reductase (EC 1.6.4.2), which catalyzes the reaction:

$$GSSG + NADPH + H^+ \longrightarrow 2GSH + NADP \quad (3.11)$$

This enzyme is located in the chloroplast stroma and is equally active in chloroplasts exposed to light or dark. It is a flavine adenine dinucleotide. (FAD) containing protein with a disulphide at the active site. The K_m of spinach leaf glutathione reductase for NADPH is 3 μM and for GSSG is 200 μM (Halliwell and Foyer, 1978). The equilibrium position of the reaction strongly favors reduction of GSSG and the glutathione pool of the stroma is maintained largely in the reduced form both in darkness and during illumination. This is important because GSSG is able to react with the thiol groups of certain enzymes to form mixed disulphides and thus inactivate them. In addition reduced glutathione is required in many biochemical reactions and is an essential coenzyme for a number of enzymes. Of particular note is the specific involvement of GSH with enzymes that detoxify herbicides such as atrazine, propachlor, and DDT.

The ratios of [GSH]:[GSSG] and [ascorbate]:[dehydroascorbate] in isolated chloroplasts have been reported to be very high in both light and darkness (Law et al., 1983). There is no evidence to show that either dehydroascorbate or glutathione play any part in the processes that cause the dark inactivation of enzymes such as the chloroplast fructose-1,6-bisphosphatase (FBPase). When the stromal ascorbate content is high, photosynthetically generated H_2O_2 can be reduced at rates comparable to its synthesis, by the ascorbate/glutathione cycle (Figure 3.4).

The Mehler reagent, methyl viologen, diverts electron flow to oxygen in illuminated chloroplasts and is the active agent in the "total kill," herbicide paraquat (Dodge, 1982). The addition of methyl viologen has no effect on the ascorbate concentration in the stroma of isolated chloroplasts in the dark, but upon illumination the ascorbate content declines rapidly so that after 1 min of illumination over 95% of the ascorbate is oxidized (Foyer et al., 1983). This is consistent with the view that methyl viologen accelerates the production of O_2^- and H_2O_2 to such an extent that the ascorbate–glutathione system is overtaxed and cannot re-reduce dehydroascorbate at an adequate rate to prevent total oxidation of the system. Similarly, the addition of 100-μM H_2O_2 to isolated chloroplasts causes partial oxidation of the stromal ascorbate in both darkness and light. The degree of oxidation is

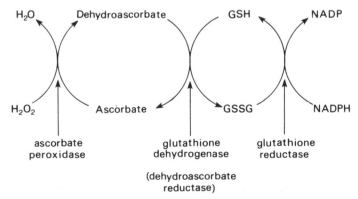

Figure 3.4. The ascorbate–glutathione cycle in chloroplasts.

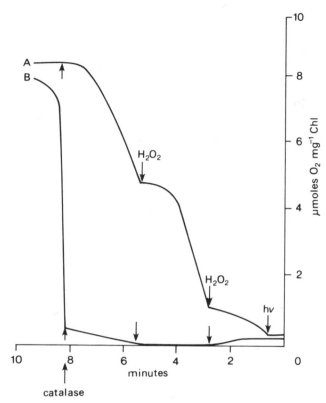

Figure 3.5. The effect of H₂O₂ on O₂ evolution by intact chloroplasts (washed 4 times in 0.5-*M* sorbitol to remove catalase) in (A) light and (B) dark. H₂O₂ was added in successive additions of 0.22-m*M* H₂O₂ as indicated, catalase (200 U/mL) was added as shown. Reaction mixtures contained 100 mg Chl/mL. The rate of H₂O₂-dependent O₂ evolution in this experiment was 48 μmol/h · mg Chl.

greater when H_2O_2 is added in the dark than when it is added during illumination, suggesting that the reduction cycle is limited in darkness by the supply of reductant, that is, NADPH (Foyer et al., 1983). The NADPH required for the glutathione reductase reaction [equation (3.11)] is obtained largely from the electron transport chain in the light and by enzymatic reactions in the stroma in the dark, such as starch breakdown. The concentration of NADPH is found to be maintained in the dark such that upon illumination only small differences in the [NADP]:[NADPH] ratio are found. Illuminated intact chloroplasts have been shown to catalyze the reduction of H_2O_2 with concomitant oxygen evolution (Figure 3.5). When ruptured chloroplasts supplied with 50-μM NADPH and substrate concentrations of GSH or GSSG are illuminated in the presence of dehydroascorbate, dehydroascorbate plus glutathione-dependent O_2 evolution has been observed with the concomitant reduction of dehydroascorbate to ascorbate (Jablonski and Anderson, 1981). Thus the series of reactions of the ascorbate–glutathione pathway for H_2O_2 removal illustrated in Figure 3.4 can be demonstrated. This cycle can provide an effective detoxification mechanism in intact chloroplasts (Figure 3.5). A similar series of reactions that appear to be inactivated in the light has been suggested to account for NADPH oxidation in the dark.

3.7.   CHANGES IN pH AND [Mg^{2+}]

The photochemical events and electron transport chain produce reducing power and the proton gradient for ADP phosphorylation. These provide the driving force for the energy requiring reactions of metabolism in the light. In the chloroplast the electron transport chain also serves to produce a favorable environment for the reactions of the RPP pathway. Photosynthetic electron transport is coupled to proton uptake into the intrathylakoid space. This establishes the proton gradient that drives photophosphorylation and causes an increase in the pH of the stroma from approximately pH 7.0 in the dark to approximately pH 8.0 in the light (Werden et al., 1975). Therefore, upon illumination the pH of the stroma is increased by almost 1 pH unit. The activity of the enzymes of RPP pathway is strongly dependent on pH such that CO_2 fixation is near the optimum at pH 8.1 but almost zero below pH 7.3. The light-driven proton uptake into the thylakoid space causes the movement of counterions, principally an uptake of chloride ions (Cl^- and an efflux of magnesium ions (Mg^{2+}) into the stroma. Thus the light-dependent alkalization of the stroma is accompanied by an increase in stroma [Mg^{2+}] (Portis and Heldt, 1976; Portis, 1981). Recent evidence suggests that light induces an increase in stromal magnesium of 2–3 mM. These Mg^{2+} movements are entirely intrachloroplastic; there is no rapid movement of extrachloroplastic Mg^{2+} through the chloroplast envelope. The total magnesium content of intact chloroplasts is from 0.4–1.0 mmole/mg chlorophyll. The concentration of free Mg^{2+} present in the stroma in the dark is still uncertain. If in the dark the free Mg^{2+} concentration is between 1 and 3 mM as has been suggested, then the light induced efflux of 20–100 $n$mole Mg^{2+}/mg chlorophyll will increase the stromal [Mg^{2+}] to between 3 and 6 mM. The remainder of the chloroplast magnesium remains chelated with the thylakoids and phosphorylated compounds. An increase in free $[Mg^{2+}]$ from 1 mM in the dark to approximately 5 mM in the light would be of sufficient magnitude to be involved in the light/dark regulation of enzymes and hence CO_2 fixation. Mg^{2+} is required for the activity of the RPP pathway, enzymes such as RuBP carboxylase, FBPases, and sedoheptulose-1,7-bisphosphatase require high $[Mg^{2+}]$ for activity. Strong inhibition of the latter two enzymes is observed when ionophores, which remove endogenous Mg^{2+} from the stroma, are added to illuminated intact chloroplasts. Consequently, CO_2 fixation is inhibited and this indicates an absolute requirement for Mg^{2+}. The changes in stromal pH and $[Mg^{2+}]$ produced by the reactions of electron transport are important in the regulation of the Calvin cycle.

3.8. REDUCTIVE ACTIVATION

While changes in pH and Mg^{2+} concentration that occur upon illumination will create optimal conditions for catalysis, several of the enzymes of the chloroplast are subject to reductive activation in the light. In particular the enzymes FBPase sedoheptulose-1,7-bisphosphatase, and ribulose-5-phosphate kinase of the RPP pathway are converted from inactive oxidized forms to reduced active forms in the light. This activation is achieved by reducing equivalents generated photochemically and transferred from the membranes to the enzymes by soluble mediators. In 1977 Wolosiuk and Buchanan demonstrated that electrons could flow from reduced ferredoxin to a sulfur-containing protein, thioredoxin, via the action of a ferredoxin–thioredoxin reductase. Ferredoxin is a low molecular weight soluble protein component of the electron transport chain. It contains two atoms of iron and two of sulfur per molecule and has an iron–sulfur group at the active site held between iron and cysteine–SH groups on the protein. Reduced ferredoxin transfers one electron per molecule to several different acceptors (Figure 3.6). The most prominent of these is NADP but other electron acceptors compete for reducing power so that ferredoxin can transfer electrons to the cyclic pathway and to oxygen in the following reactions to produce H_2O_2:

$$\text{ferredoxin reduced} + O_2 \longrightarrow \text{ferredoxin oxidized} + O_2$$

(3.12)

$$\text{ferredoxin reduced} + O_2^- + 2H^+ \longrightarrow H_2O_2 + \text{ferredoxin oxidized}$$

(3.13)

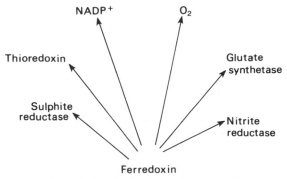

Figure 3.6. Enzymes and metabolites in the chloroplast stroma that compete for electrons from reduced ferredoxin.

Reduced ferredoxin is a cofactor for the enzymes such as sulfite reductase, nitrite reductase, and glutamate synthase [Anderson (1982) for review]. The reduction of thioredoxin, a low molecular weight hydrogen carrier protein by ferredoxin is an important mechanism of light-dependent enzyme regulation in chloroplasts (Buchanan, 1980). Reduced thioredoxin has a single catalytically active dithiol site and can act as an efficient protein disulfide reductase that reduces and thereby activates a number of regulatory enzymes in the chloroplast. The majority of the free sulfhydryl groups in chloroplasts are associated with stomal protein in nondenaturing conditions (Slovacek and Vaughn, 1982). The sulfhydryl content increases by 45–50% upon illumination and correlates with the reductive activation of FBPase under optimal conditions of pH and $[Mg^{2+}]$. The presence of DCMU, which inhibits noncyclic electron flow, has been shown to inhibit both processes. Thioredoxin is not the only mediator of thiol modulation, membrane bound reductants [designated light-effect mediators (LEMs)] that interact with the electron transport chain have been suggested to fulfill a similar function (Anderson, 1979). The LEM components would occur mainly in the oxidized (disulfide) state in the dark and in the reduced (sulfhydryl) state in the light and react with soluble enzymes in a similar manner to thioredoxin (Anderson et al., 1980). Reversal of the thiol-modulation process could account for deactivation in the dark. The ferredoxin–thioredoxin system is readily reversible and therefore could catalyze both reduction and oxidation of enzymes depending on the redox state of the system. GSSG and dehydroascorbate have been suggested to be responsible for dark inactivation of enzymes such as FBPase. However, Halliwell and Foyer (1978) have shown that glutathione is unlikely to be the physiological oxidant. Since the principal sites of inhibition of the RPP pathway by H_2O_2 are the fructose and sedoheptulose bisphosphatases, it is likely that H_2O_2 might be an effective oxidant of these enzymes *in vivo*. It has been suggested that reducing systems such as the thioredoxin system exist to ensure that the sulfydryl groups of the active forms of the bisphosphatase enzymes are kept in the reduced state despite the presence of any H_2O_2 that might escape the action of the ascorbate–glutathione system. Since H_2O_2 is generated during illumination rather than in the dark, it seems unlikely that H_2O_2 could serve as a mechanism for the prolonged inactivation of these enzymes in the dark. However, H_2O_2 could be generated in the dark by a back flow of electrons from NADPH to ferredoxin, which could then autoxidize as in equations (3.12) and (3.13). Glucose-6-phosphate dehydrogenase (EC 1.1.1.49) has been shown to be activated in the dark by an oxidative process that can be mediated by oxidized thioredoxin. Glucose-6-phosphate dehydrogenase is the first enzyme of the oxidative pentose phosphate pathway that appears to be largely inoperative in dark adapted chloroplasts, leaves, and algal cells, but is rapidly activated when the light is switched on.

3.9.   ADENYLATE SYSTEM OF THE CHLOROPLAST

In full sunlight the reactions of the RPP pathway may limit the rate of photosynthesis. In the chloroplast feedback mechanisms operate between the stromal and thylakoid reactions that determine the rate of electron transport and, through the availability of NADP, the pattern of electron flow. The total NADP and NADPH pool of the stroma is between 10 and 20 $n$mole/ mg chlorophyll. Up to 20% of this total is found in the reduced state in the dark, rising to between 40 and 50% of total in the light. Additional NADP has been shown to be synthesized from NAD in the stroma upon illumination. The enzyme that catalyzes NADP reduction, ferredoxin–NADP reductase, is a tightly bound component of the thylakoid membrane situated in partially protected sites very near to CF_1. Light induces a conformational change in the enzyme, via the formation of ΔpH, which increases its affinity for its substrates, ferredoxin and NADP.

Photophosphorylation is coupled to electron flow via a common intermediate, the electrochemical proton gradient. The electrochemical potential of the proton gradient is approximately pH 5.0 at the base of the proton channel CF_0 and pH 8.0 on the stromal side of CF_1. The formation of this proton gradient across CF_1 causes loss of bound ADP, activation of coupling factor and ATP synthesis. Superimposed upon this reversible process of pH activation is thiol modulation of CF_0–CF_1. Reduced dithiol compounds such as thioredoxin can interact with the pH activated CF_1–CF_0 complex such that the reduced enzyme is capable of hydrolyzing ATP at high rates. The CF_0–CF_1 complex is activated by ΔpH but in the oxidized state will not hydrolyze ATP at a significant rate. In intact chloroplasts both pH activation and thiol modulation operate on CF_0–CF_1 so that ATP hydrolysis can be observed when the organelles are disrupted. Upon returning to darkness the process of pH activation and thiol modulation are reversed so that after 10 min of darkness the ATPase is deactivated in intact chloroplasts (Mills and Mitchell, 1982). The oxidative system that reverses thiol modulation of CF_0–CF_1 is not as yet apparent. Thiol modulation of CF_0–CF_1 enhances ATPase activity but does not inhibit ATP synthesis from ADP and Pi. The hydrolysis of ADP is coupled to the inward translocation of protons. Uncouplers serve to stimulate the ATPase activity by dissipating the proton gradient.

The measurable concentrations of ADP and ATP in intact chloroplasts are variable. The total adenylate pool (ATP + ADP + AMP), which is approximately 40–60 $n$mole adenylates per mg chlorophyll, is kept rigidly to the equilibrium position of the adenylate kinase reaction [equation (3.14)], which is about 0.45 (Giersh et al., 1980):

$$2ADP \; \xrightleftharpoons{\hspace{2cm}} \; ATP \; + \; AMP \qquad (3.14)$$

Chloroplasts and mitochondria are actively engaged in energy conservation and both are capable of synthesizing ATP under a very high phosphorylation potential. The phosphorylation potential (P) is a measure of the energy state of the adenylate system and its ability to transfer phosphate:

$$P \; = \; [ATP]/[ADP] \; + \; [Pi] \qquad (3.15)$$

Kraayenhof (1969) reported that illuminated thylakoids would synthesize ATP up to a phosphorylation potential of 30,000 M^{-1}. At physiological concentrations of orthophosphate ($\sim$5 mM) the [ATP]:[ADP] ratio would be approximately 150 at this phosphorylation potential and the concentration of AMP would be extremely low. However the [ATP]:[ADP] ratios measured in the light in intact chloroplasts were much lower than those found with broken chloroplasts even when electron transport was under photosynthetic control. At maximum ATP levels the ADP concentration in illuminated intact chloroplast remains remarkably high, rarely falling below 0.4 mM. The [ATP]:[ADP] ratios observed in illuminated chloroplasts vary but are commonly from 1–4. This corresponds to an energy charge of about 0.8, which Atkinson (1968) suggested to a region of optimal regulation and interaction between energy-yielding and energy-consuming reactions.

In chloroplasts the envelope presents a barrier to the movement of adenylates and direct ATP translocation across the envelope is probably insignificant in mature leaves. This explains the difference between the high phosphorylation potential obtained with thylakoids (30,000 M^{-1}) and that obtained with intact chloroplasts (60–1000 M^{-1}) as the presence of the envelope prevents exchange of adenylates between the stroma and the suspending medium. In Chapter 1 it was shown that the PMF is the driving force for endergonic ATP synthesis. The phosphorylation potential built up by isolated chloroplasts in the light, when ADP stimulates electron flow and an increased [ATP]:[ADP] ratio suppresses the rate of electron transport, should be expected to be not far from equilibrium with the PMF. However with intact chloroplasts, the PMF and the phosphorylation potential have been shown to be far from equilibrium (Giersh et al., 1980). This would imply that ATP synthesis is not a rate-limiting step in photosynthesis.

High rates of photosynthesis can be observed at very low [ATP]:[ADP] ratios because the chloroplast is able to adjust its metabolism to offset a low [ATP]:[ADP] ratio by increased [NADPH]:[NADP] ratios or an increased phosphogylcerate concentration. In the RPP pathway the reduction of 3-phosphoglycerate (PGA) to dihydroxyacetone phosphate is catalyzed by the enzymes phosphoglycerate kinase, NADP-glyceraldehyde-3-*P* dehydrogenase, and triose phosphate isomerase. As long as metabolic fluxes are small the high activities of these enzymes in the stroma ensure that the reaction is not far from thermodynamic equilibrium:

$$\frac{[ATP^{4-}]\,[NADPH]\,[H^+]\,[PGA^{3-}]}{[ADP^{3-}][Pi^{2-}][NADP^+][DHAP^{2-}]} = K = 7.25 \times 10^{-7} \quad (3.16)$$

Phosphoglycerate reduction can be driven upon illumination either by an increased [NADPH]:[NADP] ratio or by an increased phosphorylation potential or both. The binding of adenylates, NADP and NADPH to proteins and other chloroplast components is not extensive, and accurate estimates of the concentrations of these metabolites *in situ* can be made. Such calculations indicate that the NADP pool is never fully reduced in the chloroplast.

In the dark the [ATP]:[ADP] in chloroplasts is low and measurements show it to be rather variable, ranging from 0.1–1.0. ATP may be synthesized in the dark from triose phosphate that is produced during starch breakdown. Chloroplasts in leaves appear to maintain higher [ATP]:[ADP] ratios in the dark than isolated chloroplasts because the chloroplast adenylate pool is linked *in situ* to the adenylate system of the cytosol.

REFERENCES

Allen, J. F. (1977a). *Curr. Adv. Plant Sci.* **29**, 459–469.

Allen, J. R. (1977b). *FEBS Lett.* **84**, 221–224.

Anderson, L. E. (1979). In *Encyclopedia of Plant Physiology* (M. Gibbs and E. Latzko, eds.), New Series, Vol. V, Photosynthesis I, pp. 271–281. Springer-Verlag, Heidelberg.

Anderson, L. E., Ashton, A. R., Bassat, B., Mohamed A. H., and Scheibe, R. (1980). *What's New in Plant Physiology* **11**, 37–40.

Anderson, J. W. (1981). In *Biochemistry of Plants* (M. D. Hatch and N. K. Boardman, eds.), Vol. 8, Photosynthesis, pp. 473–500. Academic Press, New York.

Arnon, D. I. and Chain, R. K. (1977). In *Photosynthetic Organelles: Structure and Function.* Special Issue of Plant Cell Physiol. (S. Miyachi, S. Katoh, Y. Fujita, and K. Shibata, eds.), pp. 129–147.

Asada, K. and Nakano, Y. (1978). *Photochem. Photobiol.* **28**, 917–920.

Atkinson, D. E. (1968). *Biochemistry* **7**, 4030–4034.

Bendall, D. S. (1982). *Biochim. Biophys. Acta* **683**, 119–151.

Buchanan, B. B. (1980). *Ann. Rev. Plant Physiol.* **31**, 341–374.

Chapman, K. S. R., Berry, J. A., and Hatch, M. D. (1980). *Arch. Biochem. Biophys.* **202**, 330–341.

Charles, A. and Halliwell, B. (1980). *Biochem. J.* **189**, 373–376.

Crowther, D. and Hind, G. (1981). In *Chemiosmotic Proton Circuits in Biological Membranes* (V. P. Skulachev and P. C. Hinkle, eds.), pp. 245–258. Addison-Wesley, Reading, Mass.

Dodge, A. D. (1982) *Biochem. Soc. Trans.* **10**, 73–75.

Egneus, H., Heber, U., Malthiesen, U., and Kirk, M. R. (1975). *Biochim. Biophys. Acta* **408**, 252–268.

Elstner, E. F. (1982). *Ann. Rev. Plant Physiol.* **33**, 73–96.

Foyer, C. H. and Halliwell, B. (1976). *Planta* **133**, 21–25.

Foyer, C. H., Rowell, J., and Walker, D. A. (1983). *Planta* **157**, 239–244.

Fridovich, I. (1978). *Science* **201**, 875–880.

Gierch, C., Heber, U., Kobayashi, Y., Inoue, Y. Shibata, K., and Heldt, H. W. (1980). *Biochim. Biophys. Acta* **590**, 59–73.

Groden, D. and Beck, E. (1979). *Biochim. Biophys. Acta* **546**, 426–435.

Halliwell, B. and Foyer, C. H. (1978). *Planta* **139**, 9–17.

Heber, U., Kirk, M. R., and Boardman, N. K. (1979). *Biochim. Biophys. Acta* **546**, 292–306.

Heldt, H. W., Chon, C. J., Lilley, R. McC., and Portis, A. J. (1977). In *Photosynthesis 1977* (D. O. Hall, J. Coombs, and T. W. Goodwin, eds.), pp. 469–478. Biochemical Society, London.

Hill, R. (1939). *Proc. R. Soc. Lond.* **B127**, 192–210.

Jablonski, P. P. and Anderson, J. W. (1981). *Plant Physiol.* **67**, 1239–1244.

Jablonski, P. P. and Anderson, J. W. (1982). *Plant Physiol.* **69**, 1407–1413.

Jackson, C., Dench, J., Moore, A. L., Halliwell, B., Foyer, C. H., and Hall, D. O. (1978). *Eur. J. Biochem.* **91**, 339–344.

Kaiser, W. (1976). *Biochim. Biophys. Acta* **440**, 476–482.

Kelly, G. and Latzko, E. (1979). *Naturwissenschaften* **66**, S617.

Kraayenhof, R. (1969). *Biochim. Biophys. Acta* **180**, 213–215.

Law, M. Y., Charles, S. H., and Halliwell, B. (1983). *Biochem J.* **210**, 899–903.

Lien, S. and San Pietro, A. S. (1979). *FEBS Lett.* **99**, 189–193.

Leegood, R. C., Crowther, D., Walker, D. A., and Hind, G. (1983). *Biochim. Biophys. Acta* **722**, 116–126.

Marsho, T. V., Behrens, P. W., and Radmer, R. J. (1979). *Plant Physiol.* **64**, 656–659.

Mehler, A. H. (1951). *Arch. Biochem. Biophys.* **33**, 65–67.

Mills, J. D. and Mitchell, P. (1982). *Biochim. Biophys. Acta* **679**, 75–83.

Mitchell, P. (1976). *J. Theoret. Biol.* **62**, 327–367.

Patterson, C. O. P. and Myers, J. (1973). *Plant Physiol.* **51**, 104–109.

Portis, A. R. and Heldt, H. W. (1976). *Biochim. Biophys. Acta* **449**, 434–446.

Portis, A. R. Jr. (1981). *Plant Physiol.* **67**, 985–989.

Robinson, S. P. and Wiskich, J. T. (1976). *Biochim. Biophys. Acta* **440**, 131–146.

Salin, M. L. and Bridges, S. M. (1981). *Plant Physiol.* **68**, 275–278.

Salin, M. L. and Bridges, S. M. (1982). *Plant Physiol.* **69**, 161–165.

Slovacek, R. E. and Vaughn, S. (1982). *Plant Physiol.* **70**, 978–981.

Steiger, H.-M. and Beck, E. (1981). *Plant Cell Physiol.* **22**, 561–576.

Van Ginkel, G. and Brown, J. S. (1978). *FEBS Lett.* **94**, 284–286.

Werden, K., Heldt, H. W., and Milovancev, M. (1975). *Biochim. Biophys. Acta* **396**, 276–292.

Wolosiuk, R. A. and Buchanan, B. B. (1977). *Nature* **266**, 565–567.

4

THE REDUCTIVE PENTOSE PHOSPHATE PATHWAY

4.1. PRINCIPAL ASPECTS

The RPP pathway of photosynthetic CO_2 fixation comprises 13 enzyme-catalyzed reactions that reside in the hydrophilic environment of the stroma. Photosynthetic carbon assimilation by isolated chloroplasts was first demonstrated by Arnon and coworkers in 1954 (Arnon et al., 1954; Allen et al., 1955). The rates of CO_2 fixation by these chloroplast fractions were only a small percentage of those measured in intact leaves; however, even these initial experiments demonstrated that the chloroplast could reduce CO_2 to the level of carbohydrate without the aid of the cytoplasm. The introduction of isolation procedures involving brief grinding of the leaves in a sugar or sugar alcohol osmoticum followed by rapid centrifugation facilitated the isolation of chloroplasts with envelopes intact and consequent retention of stromal enzymes. This coupled to the realization that the chloroplast required either a sugar phosphate or orthophosphate (Pi) for maximal activity enabled the demonstration of CO_2 fixation at its full potential in isolated chloroplasts (Bucke et al., 1966; Jensen and Bassham, 1966). In the 1950s Calvin and colleagues elucidated the pathway of photosynthetic conversion of CO_2 to carbohydrate (Benson and Calvin, 1950; Calvin and Bassham, 1962). Benson and Calvin were the first to describe the reaction sequence of CO_2 fixation and the RPP pathway is therefore often referred to as the *Benson–Calvin cycle* or the *Calvin cycle*. The RPP pathway is the only process by which net carbon fixation can be achieved in photosynthesis.

The RPP pathway has four fundamental features: carboxylation, reduction, regeneration, and autocatalysis.

Carboxylation

In the RPP pathway CO_2 reacts with the acceptor substrate, RuBP, to yield two molecules of PGA:

$$CO_2 + H_2O + RuBP \xrightarrow{\quad Mg^{2+} \quad} PGA \qquad (4.1)$$

Reduction

PGA is converted to glycerate-1,3-bisphosphate (GBP) using ATP in the reaction:

$$PGA + ATP \longrightarrow GBP + ADP \qquad (4.2)$$

NADPH is then utilized to produce glyceraldehyde-3-phosphate (GAP) as follows:

$$GBP + NADPH + H^+ \longrightarrow GAP + NADP \qquad (4.3)$$

GAP and its isomer dihydroxyacetone phosphate (DHAP) are collectively termed triose phosphates:

$$GAP \rightleftharpoons DHAP \qquad (4.4)$$

The equilibrium position of this reaction lies to the right such that at equilibrium 22 of every 23 molecules of triose phosphate will be present as DHAP. During photosynthesis, *in vitro*, the products of carbon fixation that leave the chloroplast stroma are DHAP and glycollate. Carbon assimilation in the chloroplast can therefore be summarized by the reaction:

$$3CO_2 + 2H_2O + Pi \longrightarrow DHAP + 3O_2 \qquad (4.5)$$

Only one of the products of photosynthesis, starch is retained in the chloroplasts but its retention is temporary. In the cytoplasm the triose phosphate exported from the chloroplasts is, in many plants, converted to sucrose, which is the major end product of photosynthesis in leaves. For continued photosynthesis the triose phosphate formed must either be exported in exchange for orthophosphate or further metabolized in the chloroplast stroma. Here, in addition to being utilized in the regeneration of Calvin cycle intermediates and in starch synthesis, newly formed carbon skeletons can be taken into the biosynthetic pathways for amino acids, lipids, and a whole range of small molecules (Leech and Murphy, 1976).

The export of phosphate esters from the stroma and import of orthophosphate (Pi) are regulated (Fliege et al., 1978). Export of triose phosphate from the chloroplast must be controlled to ensure that no more than one-sixth of the total carbon in the cycle is lost from the stroma. The inner envelope membrane does not permit the passive movement of either Pi or phosphate ester. Transfer is mediated by the phosphate translocator, which ensures an obligatory anion counterexchange such that the sum of Pi and phosphorylated intermediates within the stroma remains constant. In the steady state, five molecules of triose phosphate or phosphoglycerate out of six newly formed are required for the internal regeneration of RuBP. If the ration drops below 5 the chloroplast is depleted of RuBP and photosynthesis is inhibited. When the external supply of Pi is limited, sugar phosphate export is diminished and surplus carbon in the chloroplast is converted into starch (Heldt et al., 1977). Starch accumulation therefore commonly occurs during high intensity illumination of leaves when carbon fixation by the chloroplast exceeds the capacity of sucrose synthesis to liberate Pi in the cytoplasm, or when isolated chloroplasts are illuminated at low external concentrations of Pi. When isolated chloroplasts are suspended in a medium containing high Pi, illumination does not result in an autocatylic buildup of intermediates and resultant acceleration of oxygen evolution because the ratio of triose phosphate utilized in regeneration to that exported falls below 5 (Figure 4.1). The rate of carbon fixation is diminished because insufficient ribulose bisphosphate is formed, substrate regeneration is prevented, and the overall rate of photosynthesis is lowered. Secondary effects including starch mobilization and oxidative inactivation of enzymes due to substrate deficiency may follow. The inhibition by high Pi can be relieved by the addition of triose phosphate or phosphoglycerate to the medium. The phosphate translocator thus regulates the stromal concentrations of dihydroxy-acetone phosphate, phosphoglycerate, and Pi. In leaf cells and protoplasts triose phosphate is exported against a concentration gradient during steady-state photosynthesis. The triose phosphate concentration of the cytoplasm is considerably higher than in the chloroplast stroma. The strict counter-exchange imposed by the phosphate translocator means that export of triose phosphate may be coupled to a converse gradient of Pi.

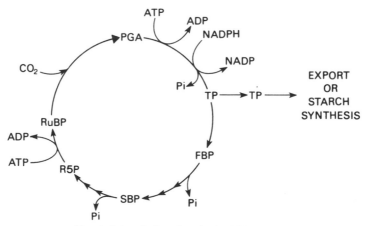

Figure 4.1. Carbon flow in the RPP pathway.

Regeneration

For every six molecules of triose phosphate formed five are rearranged in an array of complex interconversions involving nine enzymes to regenerate three molecules of RuBP, as substrate for the cycle to continue. These reactions are

$$GAP + DHAP \rightleftharpoons \text{fructose-1,6-bisphosphate (FBP)} \quad (4.6)$$

$$FBP + 3H_2O \longrightarrow \text{fructose-6-phosphate} + \text{orthophosphate} \quad (4.7)$$

$$\text{Fructose-6-phosphate} + GAP \rightleftharpoons$$
$$\text{xylulose-5-phosphate} + \text{erythrose-4-phosphate} \quad (4.8)$$

$$\text{Erythrose 4-phosphate} + DHAP \rightleftharpoons \text{sedoheptulose-1,7-bisphosphate} \quad (4.9)$$

$$\text{Sedoheptulose-1,7-bisphosphate} + H_2O \longrightarrow$$
$$\text{sedoheptulose-7-phosphate} + \text{orthophosphate} \quad (4.10)$$

$$\text{Sedoheptulose-7-phosphate} + GAP \rightleftharpoons$$
$$\text{ribose-5-phosphate} + \text{xylulose-5-phosphate} \quad (4.11)$$

$$\text{Ribose-5-phosphate} \rightleftharpoons \text{ribulose-5-phosphate} \quad (4.12)$$

$$\text{Xylulose-5-phosphate} \rightleftharpoons \text{ribulose-5-phosphate} \quad (4.13)$$

$$\text{Ribulose-5-phosphate} + ATP \longrightarrow RuBP + ADP \quad (4.14)$$

Autocatalysis

The regeneration formation of RuBP leads to the "autocatalytic," build up of pathway intermediates. This cyclic process makes the pathway self-sufficient in that it can produce and increase its own substrate (Figure 4.2).

4.2. INDUCTION

When leaves or chloroplasts are illuminated after a period of darkness, photosynthetic carbon assimilation does not commence immediately as does electron transport. There is a "lag" or "induction" period that can last for

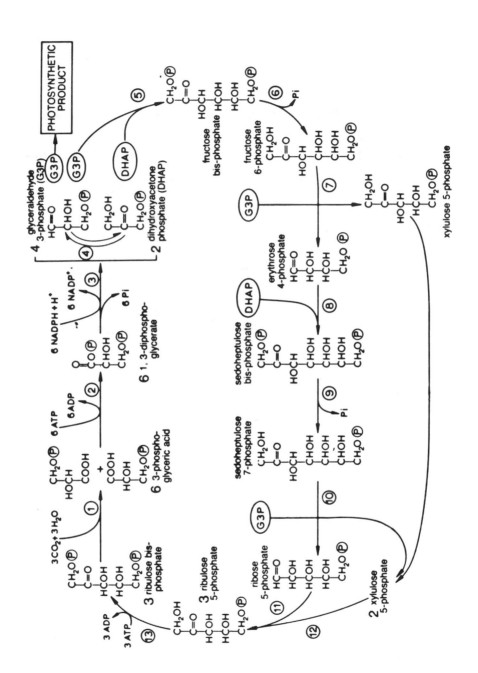

84

several minutes during which CO_2 fixation begins slowly and then accelerates until a steady-state rate is achieved (Figure 4.3). Osterhout and Haas (1918) attributed the lag phase of photosynthesis to two possible factors. Firstly, an increase in the concentration of key intermediates and, secondly, the activation of component catalysts.

Induction is a fundamental feature of photosynthesis that reflects the autocatalytic nature of RPP cycle. The RPP pathway produces its own substrate that increases, to a maximum, as the number of turns of the cycle increases. The rate of the cycle increases with increasing substrate until a maximum flux is attained. The duration of the lag phase is essentially independent of light intensity but dependent on temperature. In high light and saturating CO_2 the rate of whole plant photosynthesis is directly proportional to temperature. At low temperature the maximum rate of photosynthesis is high yet the induction period is extended when compared to higher temperatures. As a result of the autocatalytic nature of the cycle, the RPP pathway does not obey Arrhenius' law and Q_{10} values are very high at low temperatures.

During steady-state photosynthesis one-sixth of the carbon fixed is available for export while five-sixths is retained for regeneration of RuBP. However, during photosynthetic induction substantially more is required for regeneration and autocatalysis since the substrate pools of the Calvin cycle must attain the necessary concentrations for CO_2 fixation to proceed. Therefore, somewhat more than five-sixths of the photosynthate must be retained in the stroma during this period. When isolated intact chloroplasts are illuminated in the presence of high concentrations of orthophosphate (Pi), then enforced export of phosphorylated pathway intermediates through the phosphate translocator leads to the breakdown of substrate regeneration and consequently increases the lag period and lowers the rate of photosynthesis. (The strict counter-exchange imposed by the phosphate translocator ensures that the sum of Pi and phosphorylated intermediates within the stroma remain constant.) When darkened spinach protoplasts are illuminated there is a rapid rise in the ratio

Figure 4.2.   The RPP pathway showing the fate of the six triose-phosphate molecules synthesized during the photoreduction of three molecules of CO_2. The enzymes involved are (1) ribulose biphosphate carboxylase, (2) phosphoglycerate kinase, (3) NADP–GAP dehydrogenase, (4) triose-phosphate isomerase, (5) aldolase, (6) fructose bisphosphatase, (7) transketolase, (8) aldolase, (9) sedoheptulose-1,7-bisphosphatase, (10) transketolase, (11) ribose-5-phosphate isomerase, (12) ribulose-5-phosphate epimerase, and (13) phosphoribulokinase. (Taken from Reid, R. A. and Leech, R. M. (1980). *Biochemistry and Structure of Cell Organelles*, p. 67. Blackie, Glasgow.)

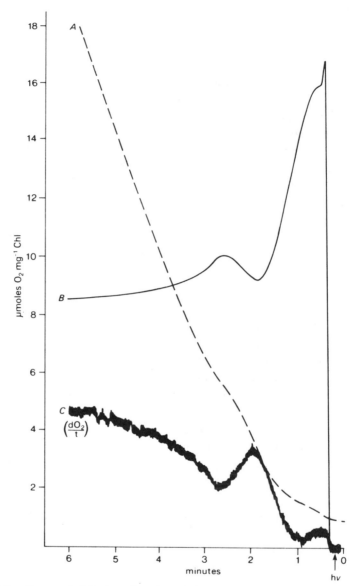

Figure 4.3. Changes in CO_2-dependent O_2 evolution and chlorophyll fluorescence associated with the lag phase of photosynthesis in barley leaf protoplasts illuminated with blue light at 120 W/m^2 showing (A) the rate of O_2 evolution, (B) chlorophyll fluorescence at 695 nm, and (C) the differentiated O_2 evolution (dO_2/T) trace. The figure shows the oscillations in fluorescence and O_2 evolution that are characteristic of the lag phase in leaves and leaf protoplasts.

of triose phosphate to PGA [reactions (4.2) and (4.3)], which reaches a maximum in the lag phase. When PGA accumulates, reduction to triose phosphate is favored and thus the rate of cycle turnover increases until a maximum rate is attained. In addition to the process of autocatalysis, the activity of the Calvin cycle is stimulated upon illumination by the light-mediated activation of several of the component enzymes. The enzymes FBPase, sedoheptulose-1,7-bisphosphatase, and ribulose-5-phosphate kinase are converted from virtually inactive to active forms by light generated reductants and by the changes in pH and Mg^{2+} concentrations that occur in the stroma upon illumination (Buchanan, 1980; Laing et al., 1981).

4.3.   THE RECONSTITUTED CHLOROPLAST SYSTEM

A reconstituted chloroplast system is one in which isolated thylakoids are reconstituted with the enzymes of the stroma and metabolites in such a way that the resultant reaction mixture is capable of catalyzing all or part of the RPP pathway when illuminated. Thus the system acts in a manner similar to intact chloroplasts except that the limiting envelope has been removed and compounds that would not normally obtain entry to the stroma across the envelope can be added so that concentrations and ratios of metabolites can be altered as required (Lilley and Walker, 1979). In the reconstituted chloroplast system there is an initial rapid burst of O_2 evolution when the light is switched on (Figure 4.4). This is caused by the reduction of NADP in the system by noncyclic electron flow, which generates ATP. When the NADP pool becomes largely reduced, noncyclic electron transport ceases but ATP can still be formed via cyclic and pseudocyclic electron flow. Substantial rates of substrate-dependent O_2 evolution can be observed upon the addition of PGA, R5P, FBP, and so on as this ultimately causes the regeneration of NADP via the activities of phosphoglycerate kinase and NADP-dependent GAP dehydrogenase [reactions (4.2) and (4.3)]. PGA-dependent O_2 evolution can be inhibited by the addition of ADP probably caused by a mass action effect on phosphoglycerate kinase (see Section 4.5). The addition of ribose-5-phosphate also stops oxygen evolution (Figure 4.4) because phosphoribulokinase acts as an effective sink for ATP and thus causes a decrease in the ratio of ATP to ADP, which, via mass action, stops NADP regeneration and hence noncyclic electron flow until such time as cyclic and pseudocyclic electron flow can restore the ATP concentration to

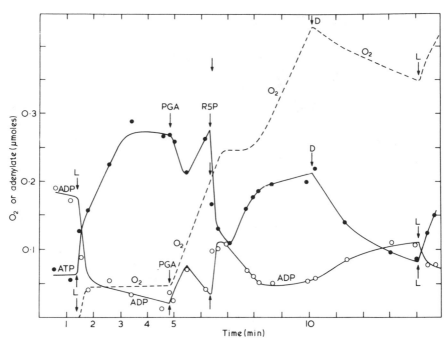

Figure 4.4. Changes in adenylate levels and O_2 evolution in the reconstituted chloroplast system upon illumination (L) and darkness (D) initially in the absence of substrate, then upon the addition of PGA, and subsequently ribose-5-phosphate (R5P). ADP (○), ATP (●), O_2 evolution (- - - - -). (Taken from Carver, K. A., Hope, A. B., and Walker, D. A. (1983) *Biochem J.* **210**, 273–276).

one that would again support PGA reduction. Photosynthetic carbon assimilation and electron transport can therefore be regulated by the controlling effects of the adenine nucleotide status (Carver et al., 1983).

4.4.   RuBP CARBOXYLASE/OXYGENASE

RuBP carboxylase/oxygenase (EC 4.1.1.39) catalyses the primary reaction of carbon fixation in the RPP pathway [reaction (4.1)] and of the photorespiratory pathway by incorporation of oxygen into RuBP. This enzyme was first purified by Wildman and Bonner (1947) from spinach leaves and called (by them) *fraction 1 protein*. In the higher plant chloroplast RuBP carboxylase/oxygenase (EC 4.1.1.3.9) is present at a protein concentration of approximately 300 mg/ml and can account for up to 65% of the total soluble protein in aqueous extracts of leaves, yet plants apparently synthesize only sufficient RuBP carboxylase/oxygenase to meet their photosynthetic requirements. The CO_2 concentration within the chloroplast during steady-state photosynthesis at 25°C is probably less than 10 μM, while the apparent K_m (CO_2) in air for a typical C3 plant is approximately 20 μM (Yeoh et al., 1981). The maximum net rate of carboxylation in this situation 1 mg CO_2/h • mg carboxylase enzyme protein present. A typical net rate of photosynthesis for a C3 plant such as wheat of 25 mg CO_2/dm^2 • h would require, therefore, 25 mg of RuBP carboxylase in this area of leaf. There is a broad correlation between RuBP carboxylase activity measured in leaf extracts and the light-saturated rate of photosynthesis. Shade leaves have low rates of light-saturated photosynthesis and smaller amounts of RuBP carboxylase than sun leaves do. Leaves of C4 plants generally contain less RuBP carboxylase than C3 plants presumably because less is required as C4 plants contain a mechanism for CO_2 concentration in the bundle sheath cells. The enzyme from C3 and crassulacean acid metabolism (CAM) species has a higher affinity for CO_2 (K_m = 12–25 μM) than does that from C4 species (K_m = 28–34 μM). The lower affinity enzymes of the C4 plants also differ from the C3 enzymes in that high inorganic carbon concentrations (> 10 mM pH 8.0) do not inhibit them, they may therefore remain fully active at the high CO_2 concentrations that obtain in the bundle sheath cells. The chloroplast RuBP carboxylase/oxygenase is freely soluble in the stroma but is often found associated by electrostatic forces to the thylakoid and envelope membranes. In the cyanobacteria the enzyme is absent from heterocysts, which are nitrogen fixing bodies. In the photosynthetic vegetative cells the enzyme is distributed between the soluble cytoplasm and the particulate polyhedral bodies, which are termed *carboxysomes*. The function of these is uncertain, they may be active in CO_2 fixation or may serve as enzyme storage reserves.

The enzyme is found to have a considerable potential activity in the dark in chloroplasts and protoplasts. Illumination has been shown to cause an approximately threefold increase in activity. However, not all workers have been able to demonstrate a decrease in carboxylase activity after the transition from light to darkness, that is, dark deactivation. The physiological significance of such low levels of activation upon illumination is uncertain, since the activity in darkness is already sufficient to allow for a significant rate of CO_2 fixation. The carboxylase activity of the enzyme [reaction (4.1)] involves the reaction of the RuBP with CO_2 and splitting of the C6 intermediate thus formed into two molecules of PGA (Lane and Miziorko, 1978). When

RuBP carboxylase was first purified, the affinity of the enzyme for CO_2 was found to be much lower than that estimated to be required in atmospheric conditions for photosynthesis in leaves or isolated chloroplasts. Similarly, the maximum velocity (V_{max}) was too low to account for observed rates of photosynthesis. However, it is now known that both the carboxylase and oxygenase activities are dependent upon the sequential binding of CO_2 and Mg^{2+} (Lorimer, 1981). The inactive form of the enzyme combines with a molecule of CO_2 in a rather slow, reversible, fashion. The enzyme–CO_2 complex is able to interact with Mg^{2+} to form a enzyme–CO_2–Mg complex. This second move rapid reversible step yields the active form of the enzyme.

$$\text{enzyme} + CO_2 \underset{(\text{inactive})}{\overset{\text{slow}}{\rightleftharpoons}} \text{enzyme} - CO_2 + Mg^{2+} \underset{(\text{inactive})}{\overset{\text{rapid}}{\rightleftharpoons}} \text{enzyme} - CO_2 - Mg^{2+}$$
$$(\text{active})$$

Thus the reaction with CO_2 and Mg^{2+} causes the conversion of the enzyme from an inactive to an active form. The activation process is relatively slow but is enhanced by a pH shift from pH 7.0 to 8.5 (Jensen and Bahr, 1977). A number of effectors such as PGA, 6-phosphogluconate, FBP, and NADPH have been shown to enhance the activation of the enzyme particularly at pH values below 8 (Buchanan and Schurman, 1972; Chollet and Anderson, 1976). Orthophosphate may also be involved in the activation process. RuBP carboxylase was found to be inhibited or inactivated under conditions of Pi deficiency (Heldt et al., 1978). Once fully activated the enzyme activity is not greatly increased by reducing agents such as dithiothreitol. The CO_2 required for enzyme activation may react directly with the enzyme protein to form a carbamate with the side chain amino group of a lysine residue. The catalytic and activation sites for CO_2 on the enzyme are distinct and separate. After formation of the enzyme–CO_2–Mg complex catalysis requires a second CO_2 molecule to bind to a second site on the active ternary complex. It has been clearly established that CO_2 and not HCO_3^- is the species required for both the activation process and for catalysis. Magnesium is not generally considered to be required for catalysis after activation is complete although it has been suggested that Mg^{2+} may stabilize the C6 intermediate formed during the reaction. Activation increases the maximum velocity of the reaction and increases the affinity of the enzyme of CO_2. If CO_2 and Mg^{2+} are removed, the active enzyme complex dissociates into the inactive form. A number of metabolites such as sedoheptulose-1,7-bisphosphate and particularly NADPH and 6-phosphogluconate have been shown to be effectors of the partially activated enzyme (Buchanan and Shurmann, 1973). In the presence of 6-phosphogluconate or NADPH significantly lower concentrations of the activating cofactors such as CO_2 and Mg^{2+} were found to activate the isolated wheat enzymes than in the absence of effectors (Gutteridge et al., 1982). Effectors such as Pi, that stimulate activation, are often found to inhibit catalysis.

The eukaryote RuBP carboxylase/oxygenases are invariably of high molecular weight (500,000–550,000) and can often be isolated in a highly purified state in a single step involving centrifugation of the soluble cell extract through a density gradient treatment (Ellis, 1979). Incubation with sodium dodecyl sulfate (SDS) or urea dissociates the molecule into large subunits (L: 50,000–55,000) and small subunits (S: 12,000–15,000). Some prokaryotes have a form of the enzyme that has only large subunits (e.g., *Chlorobium thiosulphatophilum*, 6L), these are designated *O* enzymes (McFadden, 1980). Most RuBP carboxylases have large and small subunits in a ratio of 1:1, which are termed *T* enzymes. These generally consist of an 8L8S quaternary structure. The L subunits bear the catalytic sites plus most of the effector binding sites. The amino acid composition indicates that the L subunits have been closely conserved in evolution. Substantial homology in primary structure, particularly in those regions around the catalytic site have been found between the L subunits of maize, spinach, and barley. However, this homology does not extend to the cysteinyl residue counterparts of the 2L enzyme from *Rhodospirillum rubrum*. The amino acid composition of the S subunits of different species shows very little homology. The function of the small subunits remains debatable. It has been proposed that they are involved in establishing the proper structure of the catalytically active holoenzyme and also that they may fulfill a regulatory role. The large and small subunits of the 8L8S eukaryote enzyme molecule are clustered in two layers perpendicular to fourfold axis of symmetry. In higher plants the L subunits are coded for and synthesized in the chloroplast, while the small subunit is coded for on nuclear genes and synthesized in the cytoplasm as a high molecular weight (~20,000) precursor. The additional 5000 molecular weight signal sequence is involved in the translocation of the S subunits across the chloroplast envelope. The removal of the extra amino acid sequences occurs on the envelope and the S subunits combine with the L subunits in the stroma to from the holoenzyme. The active enzyme cannot be formed until the S subunits are structured into the protein. The enzyme is synthesized only during leaf expansion and is neither synthesized nor turned over in mature leaves.

In the addition to the carboxylation reaction (4.1) RuBP carboxylase/ oxygenase also catalyses the addition of molecular oxygen into the RuBP molecule as follow

$$O_2 + \text{D-ribulose-1,5-bisphosphate} \xrightarrow{\text{Mg}^{2+}} \text{PGA} + \text{2-phosphoglycollate}$$
$$(4.15)$$

This reaction is the initiation of the photorespiratory pathway (Tolbert, 1980) and is referred to as the *oxygenase* activity of the enzyme. Because CO_2 and O_2 compete for the same active site on the enzyme, the rates of the two reactions are fundamentally dependent on the relative concentrations of the two gases. The presence of O_2, therefore, inhibits CO_2 fixation. The actual rates of the two reactions depend on other factors such as temperature, but under normal air (250-μM O_2, 10-μM CO_2) the carboxylase activity could be expected to be from 3–5 times greater than the oxygenase activity. The oxygenation of RuBP produces phosphoglycollate, which is converted to glycollate in the chloroplast. Glycollate is the substrate for the photo-respiratory pathway and is not further metabolized by the chloroplast from which it is exported. The photorespiratory pathway ultimately leads to the loss of CO_2 and thus reduces the total amount of CO_2 fixed by photosynthesis. Photosynthesis in the leaves of C3 plants grown at 25°C under the high light can always be increased either by increasing the CO_2 in the atmosphere or by decreasing the oxygen content to 2%. Detailed investigations of the RuBP carboxylase/oxygenase from many species have shown that both activities respond in a similar fashion when the activation state of the enzyme is altered or in the presence of inhibitors and effectors. Consequently, selective inhibition or removal of the oxygenase activity is probably not possible. However, the metal ion specificity of the carboxylase and oxygenase activities is different. When Mg^{2+} ions are replaced by Mn^{2+}, there is selective inhibition of the carboxylase activity relative to the oxygenase, therefore, differential modulation of the two activities is possible. The ratios of carboxylase:oxygenase activities measured in a range of plant and bacterial species suggest that a wide range of variation exists among photosynthetic organisms.

4.5. PHOSPHOGLYCERATE KINASE

The magnesium-dependent phosphoglycerate kinase reaction (EC 2.7.2.3) produces GBP at the expense of ATP [reaction (4.2)]. Unlike most kinases phosphoglycerate kinase catalyses a freely reversible reaction and, therefore, for the reaction to operate in the direction of diphosphoglycerate (GBP) formation a high ratio of substrates ([ATP]:[PGA]) to products ([ADP]:[GBP]) is required. In addition, the formation of GBP is the most energetically unfavorable reaction of the cycle ($\Delta F' \simeq 4.5$ kcal), which as a consequence of mass action makes this reaction particularly susceptible to inhibition by ADP. Two-thirds of the total ATP required by the cycle is utilized in this reaction. The phosphoglycerate kinase reaction is particularly sensitive to changes in the concentration of ATP, ADP, and PGA in the stroma. In the presence of NADPH, GBP is reduced to triose phosphate by the enzyme NADP–GAP dehydrogenase [reaction (4.3)]. This is the only reductive reaction of the RPP pathway. The steady-state concentration of GBP in the stroma is kept at a very low level relative to that of PGA because of the activity of the latter enzyme. This, in turn, displaces the equilibrium of the phosphoglycerate kinase reaction in favor of GBP formation. The reduction of PGA to GAP utilizes ATP and NADPH and so regenerates the ADP, NADP, and Pi required for continued electron transport. Phosphoglycerate kinase is not light activated as it is not apparently susceptible to allosteric regulation by metabolites. It has been suggested that this enzyme is regulated by energy charge (Pacold and Anderson, 1975) but the observed responses to energy charge (the extent of phosphorylation of the adenylate pool) are not distinguishable from those of mass action or end-product inhibition and do not appear to operate in an allosteric fashion.

4.6. NADP–GAP DEHYDROGENASE

The reversible reduction of GBP to GAP [reaction (4.3)] is catalyzed by a NADP-dependent dehydrogenase (EC 1.2.1.13), which is located exclusively in the chloroplast stroma. The purified enzyme is also able to utilize NADH but it is doubtful whether this activity has any physiological significance. The equilibrium position of reaction (4.3) favors GAP production and displaces the equilibrium of the preceding phosphoglycerate kinase reaction in favor of GBP formation. The reduction of GBP to GAP may be considered to be generally limited by the rate of NADP production by the electron transport chain. In chloroplasts the levels of PGA, ADP, and ATP are limited through the activities of phosphoglycerate kinase and NADP–GAP dehydrogenase such that a fall in ATP concentration is often accompanied by an increase in the concentrations of ADP and PGA. The spinach chloroplast $NADP^+$–GAP dehydrogenase is a multimeric protein of 600,000 molecular weight. It is active with either NADPH or NADH, binding 4 mole of either coenzyme per 600,000 (Yonuschot et al., 1970; Ferri et al., 1978).

The aggregation state of the enzyme may be changed *in vitro* by incubation with pyridine nucleotides, for example, a reversible dissociation into monomers of 145,000 occurs upon incubation with NADP or NADPH and is accompanied by an enhancement of the NADPH-linked activity. Conversely, incubation with NAD or GAP produces aggregation of the monomers and enhances the NADH-linked activity. The isolated enzyme can also be activated by dithiothreitol, which enhances the dissociating action of NADPH and favors the monomeric form. Removal of bound pyridine nucleotides causes a conformational change in the enzyme but does not change the aggregation state (Ferri et al., 1981). Probably both the monomeric and oligomeric forms exist in the chloroplast.

The activity of the enzymes has been shown to increase by severalfold in chloroplasts and leaves upon illumination and decrease on returning to darkness (Wolosiuk and Buchanan, 1978). The light activation is relatively slow and cannot always be demonstrated. The activation has been shown to be effected both by the thioredoxin system and by the LEM system. The latter would involve the reduction of membrane dithiols in the light followed by a sequential reduction of a soluble protein modulase (not identical with thioredoxin) and finally dithiol groups on the enzyme. Chemoautotrophic bacteria that do not evolve oxygen do not contain an NADP-linked enzyme but have instead an NAD–GAP dehydrogenase.

4.7. TRIOSE PHOSPHATE ISOMERASE AND ALDOLASE

The triose phosphates GAP and DHAP are maintained in an equilibrium of 1 molecule of GAP for every 22 molecules of DHAP [reaction (4.4)] by the action of triose phosphate isomerase (EC 5.3.1.1). The equilibrium position of this reaction lies far to the right in favor of DHAP. The chloroplast triose phosphate isomerase is apparently very similar to the cytoplasmic isoenzyme. The pea chloroplast enzyme has a K_m for GAP of 0.42 mM and a K_m for DHAP of 1.1 mM. With GAP as the substrate, competitive inhibition is found with phosphoglycollate (K_i = 15 μM), phosphoenolpyruvate (K_i = 1.3 mM), and RuBP (K_i = 0.56 mM) (Anderson, 1971).

The aldolase (EC 4.1.2.13) of chloroplasts participates in two reactions of the RPP pathway leading to the synthesis of FBP [reaction (4.6)] and also sedoheptulose-1,7-bisphosphate [reaction (4.9)]. Both activities are apparently catalyzed by the same enzyme (Murphy and Walker, 1982). The aldolases purified from some blue-green algae (e.g., *Anacystis nidulans*) and *Chromatium* require preincubation with cysteine and the presence of a divalent metal ion for activity. The aldolase from other photosynthetic organisms and higher plant chloroplasts does not require the presence of a divalent metal ion for activity. The chloroplast enzyme does not show light activation and is present at an activity far in excess of that required for CO_2 fixation. The reaction is freely reversible and does not appear to be allosterically regulated.

4.8.   FBPase

Chloroplasts contain a unique alkaline FBPase (EC 3.1.3.11) for the specific hydrolysis of FBP to fructose-6-phosphate and Pi [reaction (4.7)]. The chloroplast enzyme is unlike the animal fructose bisphosphatases and the cytosol enzyme. Its activity is not inhibited either by adenosine monophosphate (AMP) or fructose-2,6-bisphosphate. It is an oligomeric magnesium-dependent enzyme that displays sigmoidal reaction kinetics and is considered to be hysteretic. FBPase activity in leaves (Kelly et al., 1976) and isolated chloroplasts increases markedly upon illumination (Leegood and Walker, 1983). In the presence of optimal concentrations of magnesium, FBPase extracted from illuminated chloroplasts and assayed at pH 7.8 or above has an activity of over 30 times that of the enzyme extracted in the dark and assayed at pH 7.1 (Laing et al., 1981). Activities of well over 200 μmole/mg of chlorophyll/h have been measured in ruptured wheat chloroplasts under optimal conditions. Light activation increases the affinity of the enzyme for its substrates and causes a shift in the pH optimum (Baier and Latzko, 1975; Charles and Halliwell, 1980). This activation by light can be mimicked *in vitro* by the dithiol compound, dithiothreitol, which reduces two key disulfide bridges on the enzyme. Incubation with dithiothreitol decreases the K_m of the oxidized enzyme for FBP from 0.8 to 0.033 mM and decreases the K_m for Mg^{2+} from 9.0 to 2 mM. Changes in enzyme properties mediated by dithiothreitol are considered very similar to those occurring *in vivo*. A variety of intracellular reductants has been proposed to mediate FBPase activation *in vivo*. Buchanan and coworkers have shown that spinach chloroplast fructose bisphosphatase was activated in the light via soluble ferredoxin (Buchanan et al., 1967) and an additional protein factor, which was later separated into two components that were identified as thioredoxin and ferredoxin–thioredoxin reductase (Wolosiuk and Buchanan, 1977). Thus in this system photoreduced ferredoxin would reduce thioredoxin via ferredoxin–thioredoxin reductase, which would in turn reduce and thus activate enzymes such as fructose bisphosphatase (Figure 4.5). There are, at least, two forms of thioredoxin in chloroplasts, thioredoxins f and m. Thioredoxin f is the most efficient activator of fructose bisphosphatase, sedoheptulose bisphosphatase, phosphoribulokinase, and NADP–GAP dehydrogenase (Wolosiuk et al., 1979). Thioredoxin m preferentially activates NADP–malate dehydrogenase.

The thioredoxins are not the only chloroplast proteins that will reduce fructose bisphosphatase and the other light activated enzymes. Anderson and coworkers have shown that chloroplasts generate membrane bound reductants (the LEM system) that function in light activation via soluble proteins without the mediation of ferredoxin (Anderson et al., 1979). An iron–sulphur protein, feralterin, has also been implicated in fructose-bis-phosphatase activation. FBPase exists in three structurally unique forms: a dimer at above pH 8.0 in the absence of Mg^{2+} and FBP, an inactive oxidized tetramer that is stable at pH 7.5, and an active reduced tetramer that is also stable at pH 7.5. Reduction and activation of the tetramer involves a reduction of disulfide bridges and a change in conformation of the enzyme.

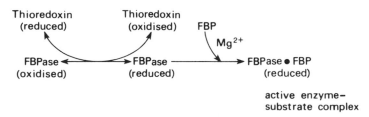

Figure 4.5. The activation sequence of FBPase.

The process by which dark deactivation is achieved is uncertain. It has been proposed that dark oxidation is brought about by GSSG. However, although GSSG does promote oxidation of the active enzyme, the glutathione pool in the chloroplast is maintained largely in the reduced form in the dark and therefore would not be effective in the process of dark deactivation (Halliwell and Foyer, 1978). Other workers have proposed that the reduction status of ferredoxin might determine the reduction status of fructose bisphosphatase and therefore could be directly involved in the dark deactivation process (Leegood and Walker, 1980). Addition of electron acceptors such as oxaloacetate to isolated chloroplasts often leads to a deactivation of fructose bisphosphatase presumably due to an oxidation of the enzyme, which could be facilitated by a direct reversal of the thioredoxin activation system.

The activation of the enzyme is stimulated by the presence of FBP. The activation of the enzyme can sometimes be inhibited in isolated chloroplasts when the external phosphate concentration is high. This is suggested to be a consequence of the depletion of the metabolite pools of the RPP pathway as discussed previously. The addition of PGA, which is known to shorten the lag phase of photosynthesis in isolated chloroplasts by stimulating the autocatalytic buildup of intermediates, increases the rate of FBPase activation.

The light activation of FBPase is inhibited in isolated chloroplasts subjected to anaerobiosis. Under anaerobic conditions inhibition of proton gradient formation will occur and this will lower the stromal pH. Consequently, a decrease in the turnover of the RPP pathway causes FBPase to remain inactive in the light. However, significant light activation of FBPase could be achieved in anaerobic chloroplasts by the addition of PGA, which causes electron transport and hence proton gradient formation as well as causing the generation of the substrate FBP. The interpretation of these findings is that FBPase but not NADP–malate dehydrogenase requires an alkaline pH and the presence of substrate for rapid reductive light activation (Leegood et al., 1983). Wolosiuk et al. (1980) have suggested that FBP and the reductant, thioredoxin, act sequentially to active FBPase as in Figure 4.5. The Mg^{2+}-independent first step of this process favors the oxidized form of the enzyme. The degree to which the enzyme will be reduced in the chloroplasts therefore depends on the electron pressure favoring reduction of the enzyme and the extent to which the reaction of the reduced enzyme with its substrate shifts the overall reaction so that the formation of the active enzyme is possible.

H_2O_2 is a potent inhibitor of FBPase, which oxidizes the activated form of the enzyme and prevents activation of the enzyme via the thioredoxin system. Thus it is possible that one function of the dithol reduction systems *in vivo* is to prevent inhibition of the reduced forms of enzymes such as FBPase by light-generated H_2O_2 (Charles and Halliwell, 1980; 1981).

4.9. SEDOHEPTULOSE-1,7-BISPHOSPHATASE

Chloroplasts contain a sedoheptulose-1,7-bisphosphatase (SBPase) (EC 3.1.3.37) that is specific for sedoheptulose-1,7-bisphosphate (SBP) [reaction (4.10)]. The enzyme has a pH optimum of approximately pH 8.2 and requires magnesium for activity. In darkness the activity of the enzyme in isolated intact chloroplasts is insufficient to support subsequent rates of photosynthesis. The activity is apparently much lower than that of the other RPP pathway enzymes. Upon illumination SBPase activity increases by about 20- to 30-fold (Laing et al., 1981). The activated enzyme has an apparent K_m for SBP of 13.5 μM. Like the FBPase, SBPase is inhibited by micromolar concentrations of Ca^{2+}. Orthophosphate competitively inhibits the enzyme activity with respect to SBP.

The kinetics of activation of SBPase suggest that this is a hysteretic enzyme. The enzyme–substrate complex is activated much more rapidly than the enzyme alone. The rate of activation *in vivo* is dependent upon the concentrations of SBP, Mg^{2+}, and photosynthetically generated reductant. As with FBPase, activation is dependent on the concentration of substrate; however, with FBPase reduction of the purified enzyme can be effected by incubation with dithiothreitol and independently of substrate activation by FBP. The reduction of SBPase does not occur until all the components (SBP, Mg^{2+}, and reductant) are present. Similarly FBPase can be activated by high Mg^{2+} and substrate in the absence of dithiothreitol but this is not the case with SBPase. If substrates are depleted (by absence of HCO_3^- or high external Pi) during illumination of intact chloroplasts then the rate of SBPase activation is severely decreased presumably because of the restriction on the SBP pool size (Woodrow and Walker, 1982). As with FBPase, thioredoxin *f* and also membrane bound dithiol groups have been implicated in the activation process.

4.10.   TRANSKETOLASE AND RIBULOSE-5-PHOSPHATE EPIMERASE

Transketolase (EC 2.2.1.1) is a glycolaldehyde transferase that reversibly transfers a 2C ketol residue from a ketose phosphate to an aldose phosphate. Thus xylulose-5-phosphate is produced by the transfer of a C2 unit from either fructose-6-phosphate (F6P) or sedoheptulose-7-phosphate (S7P) to GAP producing also erythrose-4-phosphate from the former [reaction (4.8)] and ribose-5-phosphate from the latter [reaction (4.11)]. Both the F6P- and S7P-dependent activities copurify from wheat leaves indicating that the same enzyme is responsible for both reactions. The enzyme is not light activated and closely resembles the yeast enzyme in many of its properties. The enzyme requires thiamine pyrophosphate for maximal activity. The presence of 2-mercaptoethanol or the absence of thiamine pyrophosphate promotes a dissociation of the tetrameric form of the enzyme to a dimeric form (Murphy and Walker, 1982). The dimeric form the enzyme is significantly less active than the tetramer. Magnesium stimulates the activity at alkaline pH values. Thiamine pyrophosphate and Mg^{2+} are apparently requried to form an enzyme-bound intermediate (see also Section 9.2).

Ribulose-5-phosphate epimerase (EC 5.1.3.1) catalyses the conversion of xylulose-5-phosphate to ribulose-5-phosphate [reaction (4.13)]. The reaction is freely reversible and at equilibrium the ratio of xylulose-5-phosphate to ribulose-5-phosphate is from 1–3.

4.11   RIBOSE-5-PHOSPHATE ISOMERASE AND PHOSPHORIBULOKINASE

Ribose-5-phosphate isomerase (EC 5.3.1.6) catalyses the freely reversible conversion of ribose-5-phosphate to its isomer ribulose-5-phosphate [reaction (4.12)]. Ribulose-5-phosphate is then phosphorylated using ATP to regenerate the CO_2-acceptor molecule RuBP [reaction (4.14)] via the enzyme phosphoribulokinase (EC 2.7.1.19). The equilibrium position of this reaction strongly favor RuBP formation.

The amount of phosphoribulokinase activity present in dark tissues is debatable but the enzyme has been shown to be rapidly activated upon illumination (a 400-fold activation has been demonstrated in pea chloroplasts) and rapidly deactivated on returning to darkness (Laing et al., 1981). Unlike fructose bisphosphatase and sedoheptulose bisphosphatase, phosphoribulokinase does not show a significant change in activity by a change from pH 7.0 to 8.0, nor is the rate of activation in isolated chloroplasts altered by high external orthophosphate concentration. A fourfold change in activity has been shown by changes in magnesium concentration and pH (Flugge et al., 1982) equivalent to those occurring during light/dark transitions. The maximum velocity of the enzyme is very high (approximately 1700 μmole h/mg chlorophyll) but the affinity of the enzyme for its substrates appears to be very low. The stromal concentration of ribulose-5-phosphate has been estimated to be 40 μM, yet the K_m ribulose-5-phosphate is approximately 0.22 mM. Similarly, the stromal ATP concentration has been calculated to be 0.4–1.0 mM, while the K_m ATP is 0.63 mM.

In addition to light activation the enzyme is regulated by stromal metabolites especially adenine nucleotides and the ATP:ADP ratio. The enzyme is strongly inhibited by physiological concentrations of ADP, PGA (Gardemann et al., 1982), and 6-phosphogluconate.

4.12. REGULATION

The RPP pathway is fundamentally controlled by light so that there is little or no CO_2 fixation in darkness. The enzymes of the pathway are subject to regulation by light in several ways. Upon illumination the alkalization of the stroma and the increase in the concentration of free Mg^{2+} provide optimum conditions for catalysis for many of the enzymes. Fructose bisphosphatase and sedoheptulose bisphosphatase will not readily function in the relatively acid environment of the stroma in the dark. The alkalization of the stroma is saturated at light intensities as low as 10 W/m^2. The Mg^{2+} changes are intimately related to this process and therefore show the same light-intensity dependence. These changes cannot therefore be of particular importance in the fine control of three enzymes.

The production of reductants by electron transport, which mediate light activation of some of the enzymes, are essential for the activity of the cycle. The enzymes fructose bisphosphatase, sedoheptulose bisphosphatase, and phosphoribulokinase are converted from inactive to active forms when plants are illuminated. An increase in activity of several other enzymes of the

RPP pathway has been observed upon illumination (for example, NADP–GAP dehydrogenase, ribulose bisphosphate carboxylase); however, the significance of these relatively small changes in activity, which cannot always be demonstrated, is not certain. In chloroplasts the lag phase of photosynthesis, which occurs when the light is switched on after a period of darkness, can be very short, for example, of less than 1 min duration. In this situation alkalization of the stroma, the increase in Mg^{2+} concentration, and the reductive activation of enzymes occur virtually simultaneously and it is therefore difficult to determine which factor is the most important. The light activation of enzymes shows a varying sensitivity to the addition or depletion of pathway intermediates or substrates for chloroplast metabolism. Fructose bisphosphatase and sedoheptulose bisphosphatase require both substrate and reductant at an alkaline pH to achieve maximum activity. When the lag phase of photosynthesis is several minutes long it can be shown that the duration of lag is not limited by the activation state of the enzymes alone but by the rate of autocatalytic buildup of pathway intermediates. The physiological function of light-dependent reductive activation of enzymes of the RPP pathway is still a matter for debate. Many authors have suggested that the reductive activation plays a major role in the control of CO_2 fixation by light, arguing that such a mechanism ensures that certain enzymes are virtually inoperative in the dark and fully functional in the light. This could serve to prevent the operation of futile cycles that could occur if, for example, fructose bisphosphatase and phosphofructokinase were simultaneously active. Other authors have questioned this interpretation and considered that light-dependent reductive enzyme activation may act as a repair mechanism for the reduction of oxidized sulfydryl groups and thus prevent inhibition of the active reduced forms of enzymes by light-generated oxidants such as H_2O_2 or following photoinhibitory damage. Another possibility is that this mechanism could serve as a means of enzyme regulation such that activity could be matched to the prevailing rate of photosynthesis or alternatively directly control flux through the RPP pathway. Thus light activation and dark inactivation could be a consequence of this control. The activity of an enzyme such as fructose bisphosphatase either during steady-state illumination or on returning to darkness appears to be governed principally by the competition for electrons between electron acceptors and the enzyme itself. This suggests that the redox state of the enzyme reflects the redox state of ferredoxin or a similar terminal electron carrier through which both activation and inactivation proceed. Thus the addition of an electron acceptor such as CO_2 or oxaloacetate to illuminated chloroplasts results in a decrease in the activity of enzymes such as fructose bisphosphatase especially when low rates of electron flow severely limit the generation of reductant. Light activation is a necessary property of the cycle that builds on autocatalysis and the increased enzyme activities greatly increase substrate concentrations that in turn stimulate the autocatalytic cycle.

The concentration of orthophosphate in the environment surrounding the chloroplasts has been clearly demonstrated to influence both the rate of photosynthesis (Walker, 1976) and the distribution of newly fixed carbon between export to the cytoplasm via the phosphate translocator and internal storage as starch (Steup et al., 1976). If the concentration of orthophosphate in the surrounding medium of isolated intact chloroplasts is lowered then the rate of photosynthesis declines as orthophosphate becomes limiting and substrate levels decline. The lack of external supply of orthophosphate causes the rate of [PGA]:[Pi] in the stroma to increase and this is an important stimulus for starch formation to occur as it causes allosteric activation of ADP–glucose pyrophosphorylase. Photosynthetic phosphorylation does not respond to decreasing orthophosphate until the concentration is very low such that a decline in photophosporylation is only caused when orthophosphate is virtually depleted. When orthophosphate participates in the control of photosynthetic carbon assimilation it does so via ATP. Several of the reactions of the RPP pathway have been shown to be influenced by adenylate status, either by the [ATP]:[ADP] ratio, energy charge or by simple mass action (see Sections 4.3 and 4.5). The RPP pathway requires light-generated assimilatory power (in the form of ATP and NADPH) to proceed. The generation of these in the light drives the synthesis of triose phosphate from PGA and the generation of RuBP from ribose-5-phosphate. A chloroplast that fixes CO_2 at a rate of 200 μmole/mg chlorophyll $\cdot$ h synthesizes and utilizes ATP at least 3 times that rate, which requires an extremely rapid turnover of the adenylate pool. As Pi in the stroma declines the adenylate pool will tend to shift in favor of ADP and AMP. This in turn will tend to decrease the rate of triose phosphate formation and favor an increase in PGA (see also Section 3.9). Through the mediation of Pi the requirements of the cytoplasm and more distant sinks, in the rest of the plant, can exert an influence on carbon fixation and chloroplast metabolism.

REFERENCES

Allen, M. B., Arnon, D. I., Capindale, J. B., Whatley, F. R., and Durham, L. J. (1955). *J. Am. Chem. Soc.* **77**, 4149–4155.

Anderson, L. E. (1971). *Biochim. Biophys. Acta* **235**, 237–244.

Anderson, L. E., Chin. H.-M., and Gupta, V. K. (1979). *Plant Physiol.* **64**, 491–494.

Arnon, D. I., Allen, M. B., and Whatley, F. R. (1954). *Nature* **174**, 394–396.

Baier, D. and Latzko, E. (1975). *Biochim. Biophys. Acta* **396**, 141–148.

Benson, A. A. and Calvin, M. (1950). *Ann. Rev. Plant Physiol.* **1**, 25–40.

Buchanan, B. B. (1980). *Ann. Rev. Plant Physiol.* **31**, 341–433.

Buchanan, B. B., Kalverer, P. P., and Arnon, D. I. (1967). *Biochem. Biophys. Res. Commun.* **29**, 74–79.

Buchanan, B. B. and Schurman, P. (1972). *FEBS Lett.* **23**, 157–159.

Buchanan, B. B. and Schurman, P. (1973). *J. Biol. Chem.* **248**, 4956–4964.

Bucke, C., Walker, D. A., and Baldry, C. W. (1966). *Biochem. J.* **101**, 636–641.

Carver, K. A., Hope, A. B., and Walker, D. A. (1983). *Biochem. J.* **210**, 273–276.

Calvin, M. and Bassham, J. A. (1962). In *The Photosynthesis of Carbon Compounds*, pp. 1–127. Benjamin, New York.

Charles, S. A. and Halliwell, B. (1980). *Biochem. J.* **185**, 689–693.

Charles, S. A. and Halliwell, B. (1981). *Planta* **151**, 242–246.

Chollet, R. and Anderson, L. E. (1976). *Arch. Biochem. Biophys.* **176**, 344–351.

Edwards, G. E. and Walker, D. A. (1982). In *C3, C4 Mechanisms, Cellular and Environmental Regulation of Photosynthesis*, pp. 1–542. Blackwell, Oxford.

Ellis, R. J. (1979). *Trend. Biochem. Sci.* **4**, 241–244.

Ferri, G., Comerio, G., Iadarola, P., Zappori, M. C., and Speranza, M. L. (1978). *Biochim. Biophys. Acta* **522**, 19–31.

Ferri, G., Iadarola, P., and Zapponi, M. C. (1981). *Biochim. Biophys. Acta* **660**, 325–332.

Fliege, R., Flugge, U. I., Werdan, K., and Heldt, H. W. (1978). *Biochim. Biophys. Acta* **502**, 532–547.

Flugge, U. I., Stitt, M., Freisl, M., and Heldt, H. W. (1982). *Plant Physiol.* **69**, 263–267.

Gardemann, A., Stitt, M., and Heldt, H. W. (1982). *FEBS Lett.* **137**, 213–216.

Gutteridge, S., Parry, M. A. J., and Schmidt, C. N. G. (1982). *Eur. J. Biochem.* **126**, 597–602.

Halliwell, B. and Charles, S. A. (1980). *Biochem. J.* **189**, 373–376.

Halliwell, B. and Foyer, C. H. (1978). *Planta* **139**, 9–17.

Heldt, H. W., Chon, C. J., and Lorimer, G. H. (1978). *FEBS Lett.* **92**, 234–240.

Heldt, H. W., Chon, C. J., Maronde, D., Herold, A., Stankovic, Z. S., Walker, D. A., Kraminer, A., Kirk, M. R., and Heber, U. (1977). *Plant Physiol.* **59**, 1146–1155.

Jensen, R. G. and Bahr, J. T. (1977). *Ann. Rev. Plant Physiol.* **28**, 379–400.

Jensen, R. G. and Bassham, J. A. (1966). *Proc. Natl. Acad. Sci. USA* **56**, 1095–1101.

Kelly, G. J., Zimmermann, C., and Latzko, E. (1976). *Biochem. Biophys. Res. Commun.* **70**, 193–199.

Laing, W. A., Stitt, M., and Heldt, H. W. (1981). *Biochim. Biophys. Acta* **637**, 348–359.

Lane, M. D. and Miziorko, H. M. (1978). In *Photosynthetic Carbon Assimilation* (H. W. Siegelman and G. Hind, eds.), Basic Life Sciences, Vol. II, pp. 19–40. Plenum Press, New York.

Leech, R. M. and Murphy, D. J. (1976). In *The Intact Chloroplast* (J. Barber, ed.), Topics in Photosynthesis, Vol. 1, pp. 365–401. Elsevier/North-Holland.

Leegood, R. C., Kobayashi, Y., Neimanis, S., Walker, D. A., and Heber, U. (1982). *Biochim. Biophys. Acta* **682**, 168–178.

Leegood, R. C. and Walker, D. A. (1980). *Biochim. Biophys. Acta* **593**, 362–370.

Leegood, R. C. and Walker, D. A. (1982). *Planta* **156**, 449–456.

Lilley, R. McC. and Walker, D. A. (1979). In *Encyclopedia of Plant Physiology (New Series) Photosynthesis* (M. Gibbs and E. Latzko, eds.), Vol. II, pp. 41–52. Springer-Verlag, Berlin.

Lorimer, G. H. (1981). *Ann. Rev. Plant Physiol.* **32**, 349–383.

McFadden, B. (1980). *Acc. Chem. Res.* **13**, 394–399.

Murphy, D. J. and Walker, D. A. (1982b). *FEBS Lett.* **134**, 163–166.

Osterhout, W. J. V. and Haas, A. R. C. (1919). *J. Gen. Physiol.* **1**, 1–16.

Pacold, I. and Anderson, L. E. (1975). *Plant Physiol.* **55**, 168–171.

Steup, M., Peavey, D. G., and Gibbs, M. (1976). *Biochem. Biophys. Res. Commun.* **72**, 1554–1561.

Tolbert, N. E. (1980). In *The Biochemistry of Plants* (D. D. Davies, ed.), Vol. 2, Metabolism and Respiration, pp. 488–521. Academic Press, New York.

Walker, D. A. (1976). In *The Intact Chloroplast* (J. Barber, ed.), Vol. I, Topics in Photosynthesis, pp. 235–278. Elsevier, North Holland.

Walker, D. A. and Robinson, S. P. (1978). In *Photosynthetic Carbon Assimilation* (H. W. Siegelman and G. Hind, eds.), Basic Life Sciences, Vol II, pp. 43–59. Plenum Press, New York.

Wildman, S. G. and Bonner, J. (1947). *Arch. Biochem. Biophys.* **11**, 381–413.

Wolosiuk, R. A. and and Buchanan, B. B. (1977). *Nature* **266**, 265–267.

Wolosiuk, R. A. and Buchanan, B. B. (1978). *Plant Physiol.* **61**, 669–671.

Wolosiuk, R. A., Crawford, N. A., Yee, B. C., and Buchanan, B. B. (1977). *J. Biol. Chem.* **254**, 1627–1632.

Wolosiuk, R. A., Pevelmuter, M. E., and Cheliebar, C. (1980). *FEBS Lett.* **109**, 289–293.

Woodrow, I. E. and Walker, D. A. (1982). *Arch Biochem. Biophys.* **216**, 416–422.

Yeoh, H. H., Badger, M. R., and Watson, L. (1981). *Plant Physiol.* **67**, 1151–1155.

Yonuschot, G. R., Artwerth, B. J., and Koeppe, O. J. (1970). *J. Biol. Chem.* **245**, 4193–4198.

5

THE ROLE OF THE CHLOROPLAST ENVELOPE

5.1. STRUCTURE AND COMPOSITION

The higher plant chloroplast *in situ* is seen in electron micrographs to be bounded by two distinct membranes, together called the *chloroplast envelope* (Plate 5.1). The chloroplasts of brown algae and chromophyta are surrounded by an additional pair of membranes referred to as the *chloroplast endoplasmic reticulum*, while the chloroplasts of the Dinophyceae and Euglenophyceae have a single additional membrane. The two membranes of the higher plant chloroplast envelope generally stain with approximately equal electron density and have a total thickness of about 6 nm. The two membranes are separated by a distance of from 10–20 nm, but there are some areas where they are more widely separated and others where they converge at random intervals and there is close membrane contact.

Mackender and Leech (1970) were the first to report on an isolation procedure for the chloroplast envelope and subsequently methods have been refined such that very pure envelope membrane preparations can be obtained (Douce and Joyard, 1982). An efficient isolation preparation of envelopes from spinach chloroplasts has been described in which chloroplasts are initially purified by isopycnic centrifugation through Percoll gradient. The envelopes subsequently detached by swelling in a hypotonic buffer in the presence of Mg^{2+}. The envelope fraction can then be separated on a density gradient as the buoyant density of the envelope membrane ($d = 1.12$) is different from that of the thylakoids ($d = 1.17$). The resultant envelope preparations do not contain significant amounts of chlorophyll or cytochromes and are composed of an homogenous mixture of both outer and inner envelope

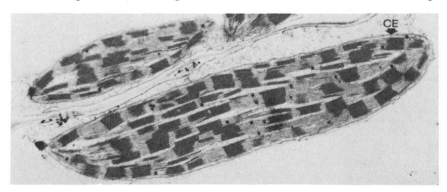

Plate 5.1. Mesophyll chloroplast from a green maize leaf ($\times$ 58,800) showing the chloroplast envelope (CE)

membranes being essentially free of contamination by other membranes (Douce and Joyard, 1979).

Separate isolation of the inner and outer envelope membranes has been achieved by rather different methods from those employed for the isolation of the membranes together (Cline et al., 1981; Block et al., 1983). When isolated intact chloroplasts are subjected to osmotic shrinkage in a hypertonic medium, an increase in the size of the intermembrane space is observed. The outer envelope membrane can then be removed by extruding these chloroplasts through a small aperture under pressure. The resultant mixture may then be subjected to sucrose density gradient centrifugation after which two envelope membrane fractions of different densities can be obtained. The lighter of these envelope membrane fractions has been shown to be enriched in the outer envelope membranes, while the heavier fraction was shown to be enriched in the inner envelope membranes (Douce and Joyard, 1983).

Together the chloroplast envelope membranes contain only 0.6–0.8% of the total chloroplast proteins. This is rather low (protein:polar lipids = 0.5–0.8) compared to the thylakoid membranes (protein:polar lipids = 2.0). Freeze-fracture cleavage of the outer and inner envelope membranes from spinach and barley has revealed four fracture faces (Van Besouw and Wintermans, 1978; 1979). The inner envelope membrane has been shown to be asymmetric while the outer envelope membrane appears to be rather similar on both faces. In spinach chloroplasts, the side of the inner membrane that is in contact with the stroma, has been found to contain particles of 70 Å at a density of approximately 1820 particles/μm^2, whereas the outer surface of this membrane has 90 Å particles at a density of 980 particles/ μm^2 (Sprey and Laetsch, 1976). The outer envelope membrane has fewer particles (130–150 particles/μm^2). The size of the particles on both the outer and inner surfaces of the outer envelope membrane is approximately 90 Å. As with the thylakoid membranes, the particles observed in freeze-fracture faces of the envelope membranes are considered to represent proteins.

The spinach envelope polypeptides have been analyzed using polyacrylamide gel electrophoresis in the presence of either SDS or lithium dodecyl sulfate (LDS). Such studies have revealed a large number of polypeptides (>75) ranging in molecular weight from 140,000 to less than 12,000. The polypeptide pattern of the envelope is clearly distinct from that of the thylakoids. There are seven major polypeptides staining intensely with coomassie blue, which together account for more than 70% of the total protein. The most prominant polypeptide has a molecular weight of approximately 29,000 and is the only polypeptide identified. Flugge and Heldt (1981) have clearly shown it to be the protein component of the phosphate translocator. The identities of the remaining major polypeptides and the large number of minor bands remain to be elucidated. Douce and Joyard (1981; 1982) have suggested that the major polypeptides are derived from proteins involved in the regulation of metabolite transport across the envelope, while the minor polypeptides are involved in other processes such as synthesis.

The separated and purified fractions of outer and inner envelope membranes show very different polypeptide patterns when analyzed by SDS–PAGE. The light fraction that is enriched in the outer envelope membrane was found to be devoid of the 29,000 polypeptide but to contain major polypeptides of molecular weights 10,000 and 24,000. While the heavy fraction that is enriched in the inner envelope membrane was found to contain the 29,000 polypeptide, it was devoid of the 10,000 and 24,000 molecular weight polypeptides, which have been localized on the outer surface of the outer envelope membrane.

Analyses of the higher plant chloroplast, amyloplast, and etioplast envelope membrane suggest that the structure of the plastid envelope has been highly conserved so that there is a large degree of uniformity in the structural organization and chemical composition of the envelope membranes of various types of plastids. The surface of the outer chloroplast envelope membrane that is in contact with the cytoplasm is strongly negatively charged. The negative charges, which are uniformly distributed over the outer membrane surface, may serve several important functions, for example, in the transport of anions and proteins and in the interactions between the outer envelope membrane and other membranes. The isolated outer envelope membrane has been found to contain the Calvin cycle enzyme RuBP carboxylase/oxygenase. However, this is most probably an artifact; as in the presence of Mg^{2+} this enzyme as well as other stromal proteins may bind unspecifically to the negative charges on the chloroplast envelope membranes when the integrity of the chloroplast is lost.

Galactolipids account for 80% of the polar lipid component of the envelope membranes. The predominant fatty acids in the polar lipids are the same as those of the thylakoids (Douce and Joyard, 1979). When the activity of galactolipid–galactolipid galactosyltransferase is lost by treatment with the proteolytic enzyme, thermolysin (Dorne et al., 1982a,b), the isolated envelope membranes were found to contain virtually no diacylglycerol. There are

two main galactolipids, MGDG and DGDG. There are two main phospholipids, PC and PG. There is also sulfolipid, which can be used as a specific
marker for plastid membranes. Qualitatively, the polar lipids of both envelope
membranes are identical, but in the outer envelope membrane the ratio of
MGDG:DGDG is 0.6:1 and that of PC:PG is 3:1, while in the inner envelope
membrane the ratio of MGDG:DGDG is 1.6:1 and PC:PG is approximately
0.5–0.8:1. In nearly all respects (except phosphatidyl choline) the polar
lipid composition of the envelope membranes is similar to that of the plasma
membrane of blue-green algae.

Plastoglobuli are not found associated with the envelope membranes.
The envelope does not contain chlorophyll but does contain carotenoid
pigments making them appear yellow in the isolated form. The carotenoid
content is distinct from that of plastoglobuli, which can act as stores for
carotenoids. Qualitatively, the carotenoid composition of the envelope is
similar to that of the thylakoid membranes (Douce et al., 1973) but the
envelope membranes are enriched in violaxanthin and depleted in β-carotene
when compared to the thylakoids. The ratio of xanthophyll to carotene has
been calculated to be approximately 6 in the envelope and approximately
3 for the thylakoids. Much of the carotenoid in the envelope is localized in
the inner membrane (10 μg/mg protein). The outer envelope membrane has
been found to contain much less carotenoid (2 μg/mg protein) and has a
much lower proportion of neoxanthin than the inner membrane. Prenylqui-
nones such as plastoquinone-9, phylloquinone, and α-tocophenol are present
in the envelope (Lichtenthaler et al., 1981) but the function of these has
yet to be made clear. There is no evidence to suggest that the photoreceptive
pigment, phytochrome, is associated with the chloroplast envelope.

5.2.   ENZYME ACTIVITIES OF THE ENVELOPE

A number of enzyme activities have been convincingly shown to be associated
with the chloroplast envelope membranes. A characteristic Mg^{2+}-dependent
ATPase is localized on the envelope. This enzyme is usually found to have
a low activity; for example, in C3 plants the activity is below 4 μmole/h ·
mg chlorophyll. The function of this enzyme is as yet uncertain; it may be
involved in protein transport or in the slow efflux of protons from the stroma
that occurs upon illumination to counteract the passive influx of protons
from the cytosol. The chloroplast envelope has also been shown to possess

a low adenylate kinase activity (Murakami and Strotman, 1978) and a high level of cyclic nucleotide phosphodiesterase (Brown et al., 1980). Both subunits of RuBP carboxylase have been found associated with isolated chloroplast envelope (Joyard et al., 1982) and this has led to the suggestion that this is of physiological significance. For example, the envelope could be the site of assembly of the two carboxylase subunits into the holoenzyme.

UDP galactose is synthesized in the cytoplasm (Leech and Murphy, 1976) and is incorporated into both MGDG and DGDG by an enzyme system localized in the chloroplast envelope (Figure 5.1). Two enzymes of the galactolipid biosynthetic pathway have been shown to be localized solely on the plastid envelope (Douce, 1974). One is UDP galactose, diacylglycerol galactosyltransferase, which catalyzes the synthesis of MGDG [reaction (5.1)] and is localized on the inner envelope of membrane:

diacylglycerol + UDP galactose ⟶ MGDG + UDP   (5.1)

The second enzyme is either a UDP galactose, MGDG galactosyltransferase,

MGDG + UDP galactose ⟶ DGDG + UDP   (5.2)

or a galactolipid, galactolipid galactosyltransferase activity,

MGDG + MGDG ⟶ DGDG + diacylglycerol   (5.3)

Treatment with thermolysin destroys the latter activity in isolated chloroplasts suggesting that this enzyme is localized on the outer surface of the outer envelope membrane and that it is not directly involved with DGDG synthesis. Dorne et al. 1982b have found that the site of [^{14}C]-galactose incorporation (from UDP-[^{14}C]-galactose) into galactolipids is located on the inner envelope

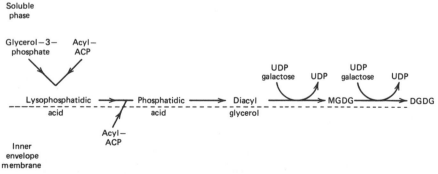

Figure 5.1. The synthesis of MGDG and DGDG from glycerol-3-phosphate and acyl-acylcarrier protein (acyl-ACP).

membrane, while Cline and Keegstra (1981) have found that MGDG synthesis occurs only on the outer envelope membrane. Clearly there is a discrepancy between the two results but this may be due to the presence of the galactolipid–galactolipid galactosyl transferase that synthesizes diacylglycerol on the outer membrane [reaction (5.3)]. Douce and Joyard (1982) suggest that any contamination of the purified outer membrane preparation by inner membrane fragments that contain the UDP–galactose diacylglyceroltransferase [reaction (5.1)] can result in the formation of high amounts of galactolipids.

Isolated chloroplasts are able to synthesize diacylglycerol by the acylation of cytoplasmically synthesized glycerol-3-phosphate followed by dephosphorylation of phosphatidic acid in the Kornberg-Pricer pathway. Joyard and Douce (1977) have shown that the enzymes involved in this series of reactions are localized on the chloroplast envelope. There are two acyl transferases associated with the pathway. The first catalyzes the synthesis of lysophosphatidic acid [reaction (5.4)] and is largely soluble. It appears to be only loosely associated with the envelope:

$$\text{glycerol-3-phosphate} \ + \ \text{acyl-ACP} \longrightarrow \text{lysophosphatidic acid} \ + \ \text{ACP} \tag{5.4}$$

where ACP represents acyl carrier protein. The second enzyme is thought to be an acyl-ACP monoacylglycerol phosphate acyl transferase catalyzing the formation of phosphatidic acid [reaction (5.5)] and is tightly bound to the envelope:

$$\text{lysophosphatidic acid} \ + \ \text{acyl-ACP} \longrightarrow \text{phosphatidic acid} \ + \ \text{ACP} \tag{5.5}$$

A specific alkaline phosphatidic acid phosphatase, which has been found to be associated with the inner envelope membrane, produces diacylglycerol:

$$\text{phosphatidic acid} \ + \ H_2O \longrightarrow n\text{-1,2-diacylglycerol} \ + \ \text{Pi} \tag{5.6}$$

Only the inner envelope membrane appears to be able to incorporate radioactivity labeled glycerol-3-phosphate into MGDG although both envelope membranes are able to synthesize phosphatidic acid.

In plant cells the plastids are the major site of fatty acid biosynthesis. These fatty acids serve as precursors for lipids synthesized in other organelles and must be transported across the envelope membrane. The long chain acyl coenzyme A (CoA) synthetase(s), which catalyze(s) reactions as shown in (5.7), are found to be present on the envelope membrane:

$$\text{fatty acid } + \text{ ATP } + \text{ CoA} \longrightarrow \text{acyl CoA } + \text{ AMP } + \text{ PPi}$$
$$(5.7)$$

Acyl-CoA synthesis can therefore be stimulated in isolated intact chloroplasts by the addition of ATP and CoA (Sanchez and Mancha, 1981).

The envelope membranes appear to have some involvement with the synthesis of carotenoids (Douce and Joyard, 1982). The enzymes responsible for the biosynthesis of carotenoid pigments from mevalonic acid (Charlton et al., 1967) and isopentenylpyrophosphate (Porter and Spurgeon, 1979) are localized within the chloroplasts. It is highly likely that there is cooperation between membrane components on the envelope and stromal enzymes in the synthesis of geranylgeraniol derivaties since the prenyltransferase complex associates with the envelope in such a way that the product of the reaction is liberated in the lipid phase of the membrane where it is further metabolized (Block et al., 1980). An enzyme system bound to the envelope has been shown to be able to catalyze the conversion of [^{14}C]-zeaxanthin into labeled violaxanthin (Costes et al., 1979). Chloroplasts contain all the enzymes required for the synthesis of α-tocopherol, phylloquinone, and plastoquinone-9 (Soll et al., 1980). It has been suggested that a α-tocopherol, phylloquinone, and most probably also plastoquinone are synthesized by enzymes associated with the plastid envelope.

5.3. PERMEABILITY

In higher plants photosynthesis is compartmentalized in the chloroplast, which is bounded by the envelope membranes that serve both as a barrier separating the chloroplast stroma from the cytosol and a bridge enabling rapid exchange of specific metabolities between the two (Heber and Walker, 1979). The outer envelope membrane is nonspecifically permeable to all molecules, both charged and uncharged, for example, sucrose and sorbitol, up to a molecular weight of approximately 8000. In contrast, the inner envelope membrane is impermeable to sucrose and sorbitol and provides an effective barrier against the unidirectional movement of most anions and cations (Figure 5.2) The inner envelope membrane is selectively permeable to certain anions because of the presence of specific translocators. However, when the term *impermeable* is applied to the inner envelope membrane, it can only be used in the restricted sense that the rate of penetration of metabolities such as sucrose is extremely low. The relative impermeability

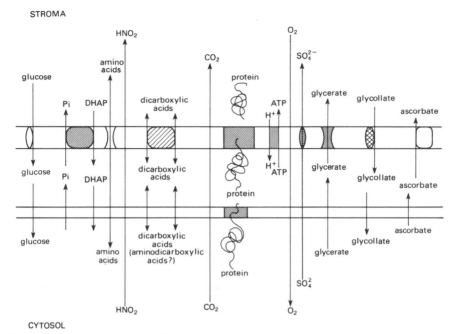

Figure 5.2.   Diagram of the main exchanges of metabolites across the chloroplast envelope in the light.

of the inner envelope to the electron acceptor, ferricyanide, provides a means of assessing the intactness of isolated chloroplast preparations (Figure 5.3). The uncoupler ammonia rapidly crosses the chloroplast envelope and will produce an almost immediate collapse of the transthylakoid pH gradient and thus uncouple electron transport from photophosphorylation (Figure 5.3).

Gases freely permeate the chloroplast envelope. The permeability of the envelope to CO_2 is very high while that of HCO_3^- is relatively lower. Thus when bicarbonate is added to isolated intact chloroplasts it is found to distribute across the envelope as would be predicted by the Henderson–Hasselbalch equation. The permeability of HCO_3^- is much lower than that of nitrite and nitrate. Undissociated acids such as propionate and acetate anions permeate more rapidly than either glycollate or glyoxalate.

At air levels of O_2 the oxygen concentration inside the illuminated chloroplast is similar to that outside. However, the O_2 concentration isolated chloroplasts during illumination has been found to be always somewhat

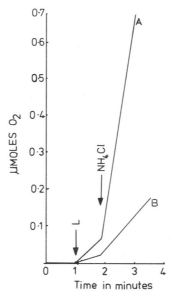

Figure 5.3. Intactness assay for isolated spinach chloroplasts measured by ferricyanide-dependent O_2 evolution in an oxygen electrode (A) ruptured chloroplasts and (B) intact chloroplasts in the absence and presence of the uncoupler NH_4Cl. The chloroplast preparation can be calculated from these measurements to be 82% intact.

higher than that of the suspending medium (Steiger et al., 1977). The degree of the gradient from the oxygen source (the thylakoids) to the sink (the suspending medium) was found to be dependent upon the O_2 concentration in the environment surrounding the chloroplasts and independent of light intensity. The ratio of the internal $[O_2]$ to the external $[O_2]$ was found to be as large as 5 at low external O_2 (30 μM) concentrations whereas at external concentrations equivalent to those in air-saturated water the ratio was close to unity. The presence of a significant O_2 gradient across the chloroplast envelope at low external O_2 is surprising since the nonpolar O_2 molecule should not be retained by biomembranes. These results suggest that the chloroplast envelope, at least in intact isolated chloroplast preparations, may represent a permeation barrier to O_2.

In addition to gas exchange discussed above, the chloroplasts of C3 plants have been shown to be primarily phosphate-importing the triose phosphate-exporting organelles (Heber and Heldt, 1981). In addition, the photorespiratory pathway in C3 chloroplasts produces glycollate, which can become a major export with glycerate (the product of the recycling glycerate pathway of photorespiration) as a significant import. The bundle sheath chloroplasts of C4 species import malate and export pyruvate in addition to PGA–triose phosphate exchange. The mesophyll chloroplasts of C4 plants primarily import pyruvate, orthophosphate, and oxaloacetate and primarily export phosphoenolpyruvate and aspartate or malate. The impermeability of the inner envelope membrane to orthophosphate, phosphate esters, dicarboxylates, and certain other metabolites is overcome by the presence of specific translocators.

5.4. TRANSPORT OF POLYPEPTIDES

Although chloroplasts carry out protein synthesis, 70–80% of the stromal proteins and thylakoid polypepides are coded for on nuclear genes and synthesized in the cytoplasm during plastid development. These proteins must then be transported into the chloroplast across the envelope membranes. The mechanism by which this is achieved is not understood although it is clear that protein synthesis and transport are independent events. The synthesis and transport of the small subunit of RuBP carboxylase/oxygenase and that of two polypeptides of the LHC protein complex have been studied (Chua and Schmidt, 1978; Highfield and Ellis, 1978). In both these cases the polypeptides are synthesized on free ribosomes in the cytoplasm as precursors of a higher molecular weight than the mature proteins. Light is not essential for the synthesis, transport, or assembly of the majority of chloroplast polypeptides but light stimulates nuclear and plastid messenger RNA accumulation and is also necessary for the synthesis of chlorophyll. The translation activity of messenger RNAs for two chloroplast polypeptides has been shown to increase following irradiation of plants with light operating through phytochrome (Apel, 1979; Tobin, 1981). The polypeptide precursors possess an additional amino terminal chain sequence of 3500–5000 molecular weight (Highfield and Ellis, 1978). When the posttranslational supernatants are incubated with purified intact chloroplasts from either pea or spinach, the pea and spinach protein precursors are found to be transported interchangeably into the chloroplasts. The uptake of the precursors has been found to be stimulated in the light suggesting that an energy-dependent process is involved. The additional amino acid sequences are removed, either during or immediately after transport across the membrane, to leave the mature proteins that can then be assembled into their required structures.

It has been suggested that the additional amino acid sequence in the precursor proteins carries the information necessary for the specific binding of the precursor to a receptor site on the envelope membrane. It is possible that the envelope protein carrier is localized in regions where the inner and outer membranes are in contact or where the inner membrane invaginates to form vesicles. The position of the receptors and the mechanism of transfer remain to be elucidated (Douce and Joyard, 1982) but it is apparent that this process does not involve a direct injection of the newly synthesized polypeptides through the envelope by membrane-bound ribosomes.

5.5.   THE pH GRADIENT ACROSS
THE CHLOROPLAST ENVELOPE

The optimal pH of the suspending medium for photosynthesis in isolated chloroplasts is generally found to be from pH 7.6–7.8. Because of the Donnan distribution of protons, the stromal pH is lower than the pH of the medium while the chloroplasts are in darkness. When isolated chloroplast suspensions are illuminated the pH of the stroma increases relative to that of the suspending medium largely because of the translocation of protons into the thylakoid space. This alkalization of the stroma is fundamentally important in the regulation of CO_2 fixation (see Section 4.12). The chloroplast envelope has a low permeability to protons but these move across the envelope membrane much more rapidly than does Mg^{2+}. If there was no mechanism to counter the passive influx of protons from the cytosol or suspending medium into the stroma, the proton gradient produced across the envelope upon illumination would slowly collapse. The pH of the stroma would become more acid and produce an environment that was not favorable to the activity of the Calvin cycle. This might be prevented by the presence on the envelope of a Mg^{2+}-dependent ATPase, which slowly pumps protons from the stroma into the medium. The net proton efflux from the stroma is compensated for at least in part by a counterflux of K^+ or Na^+ ions. This would not be required if the active proton efflux was balanced by proton influx. Loss of K^+ from the chloroplasts has been shown to result in acidification of the stroma and the inhibition of photosynthesis (Kaiser et al., 1980). Photosynthesis in isolated chloroplast preparations may be inhibited by the addition of Mg^{2+} and other divalent cations (Huber, 1978). Mg^{2+}

does not readily cross the envelope membrane, suggesting that Mg^{2+} acts directly on the membrane. The presence of Mg^{2+} has been shown to cause an acidification of the stroma that is accompanied by an efflux of K^+. External K^+ prevents this effect, which may explain why Mg^{2+} inhibition of photosynthesis is not always observed. Active ion fluxes are only observed in the light and serve to maintain the proton gradient across the envelope by counteracting the passive equilibration of ions toward the Donnan distribution. Taking the Donnan distribution of protons into account the effective ΔpH across the envelope on illumination can be calculated to be approximately 0.8. Low concentrations of carbonyl cyanide-m-chlorophenylhydrazone (CCCP) increase the proton conductivity of membranes. Similarly, the addition of low concentrations of this compound or the salts of certain weak acids, for example, nitrous acid, to isolated illuminated chloroplasts decreases or abolishes the pH gradient. For a salt to produce this effect the chloroplast envelope must allow the penetration of the undissociated acid and its corresponding anion into the stroma such that both species can operate a proton shuttle. High concentrations of bicarbonate also decrease ΔpH across the envelope. However, the permeability of the chloroplast envelope to bicarbonate is low while that of CO_2 is very high. The $HCO_3^- - CO_2$ system is therefore an ineffective shuttle for protons or hydroxyl ions at physiological concentrations.

5.6. THE PHOSPHATE TRANSLOCATOR

The inner envelope membrane is largely impermeable to anions but this membrane possesses a specific translocator protein that facilitates the counterexchange of orthophosphate, triose phosphate, and PGA. The major flow of metabolites across the chloroplast envelope is mediated by the phosphate translocator, which enables the specific transport of orthophosphate and phosphorylated compound such that photosynthetically fixed carbon in the form of triose phosphate can be exported from the stroma to the cytosol in stoichiometric exchange for orthophosphate. Export of triose phosphate via the phosphate translocator is driven by the gradient of orthophosphate that exists between the stroma and the cytosol. Studies with the pH dependence of transport has shown that metabolites are transported bound to the translocator protein as divalent anions. For each orthophosphate or phosphate ester molecule imported into the stroma, another is exported such that the total phosphate content of the chloroplast is maintained at a constant level (Figure 5.4).

The equilibrium between the trivalent phosphoglycerate ion and its divalent protonated form is pH dependent. When the pH of the stroma becomes alkaline upon illumination trivalent phosphoglycerate ion is predominant. Since the phosphate translocator transports only divalent anions, the trivalent phosphoglycerate is retained in the chloroplast stroma. In contrast, dihydroxyacetone phosphate and orthophosphate are divalent ions. Thus during CO_2 fixation dihydroxyacetone phosphate is preferentially exported from the stroma resulting in low external and high internal ratios of PGA:DHAP. This, in turn, permits efficient reduction of PGA at low NADPH:NADP

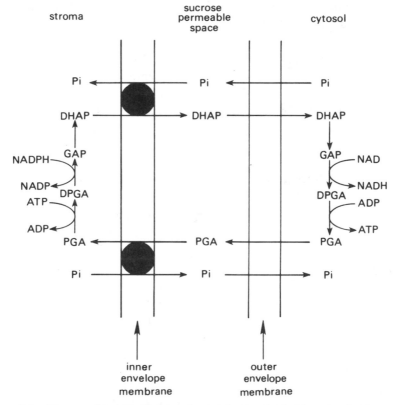

Figure 5.4. Diagram of the phosphate translocator showing the shuttle system for the transport of ATP and reducing power across the chloroplast envelope.

ratios and at low phosphorylation potentials. In this way the phosphate translocator is involved in the regulation of the stromal concentrations of dihydroxyacetone phosphate, its oxidation product PGA, and orthophosphate. The stromal ratio of PGA:Pi, in turn, determines the distribution of photosynthetic products in the light and the mobilization of starch in the dark (Chapter 6). Since PGA transport has been shown to be inhibited by Mg^{2+} then the rise in stromal Mg^{2+} that occurs upon illumination may also contribute to the retention of PGA in the stroma during CO_2 fixation. Pyrophosphate and citrate bind to the phosphate translocator but are not transported at a significant rate and are therefore useful competitive inhibitors of orthophosphate or phosphate ester transport.

A second important function of the phosphate translocator is to link the intra- and extrachloroplast pyridine nucleotide and adenylate systems. Photosynthetically produced ATP and NADPH are not directly available to the extrachloroplast compartments due to the low permeability of the inner envelope membrane to these compounds in mature tissue. The phosphate translocator provides an indirect shuttle system for transferring ATP and NADPH to the cytoplasm involving exchange of dihydroxyacetone phosphate and PGA (Figure 5.4). This shuttle will operate in either direction depending on the redox potential of the pyridine nucleotides in the cytoplasm and stroma.

Flugge and Heldt (1977; 1981) have shown that the major polypeptide of the chloroplast envelope, which has a molecular weight of approximately 29,000, is involved with the transport of orthophosphate and sugar phosphates across the inner envelope membrane. Sulfydryl reagents such as p-chloromecuriobenzoate inhibit the activity of the translocator. The 29,000 polypeptide has been shown to bind ^{35}S-p-(diazonium)-benzene sulfonic acid, a sulfydryl inhibitor of the translocator and labeling with this compound was found to be significantly decreased by the addition of substrates of the translocator to the incubation medium with the inhibitor. Reagents such as pyridoxal-5'-phosphate, which forms a Schiff base with lysine or reagents that react with arginyl residues, inhibit the activity of the translocator (Flugge and Heldt, 1977; 1978). Experiments with these compounds indicate that lysine and arginine provide the cationic groups on the protein that are involved in the binding of the transported divalent anions. Such observations led Flugge and Heldt (1977) to conclude that the 29,000 polypeptide of the envelope was involved in the activity of the phosphate translocator. In support of their conclusion Flugge and Heldt (1981) have incorporated the solubilized and purified polypeptide into liposomes and reconstituted the activity of the phosphate translocator. Flugge (1982) has shown that the phosphate translocator protein is coded for on nuclear genes.

5.7.   THE DICARBOXYLATE TRANSLOCATOR

Heldt and Rapley (1970) demonstrated that dicarboxylic acids such as oxaloacetate, malate, aspartate, and 2-oxoglutarate are transported across the inner envelope membrane by a specific translocator or translocators with overlapping specificity (Lehner and Heldt, 1978). The dicarboxylate translocator operates a counterexchange that is not strictly coupled to the simultaneous cotransport of another dicarboxylate ion. The shuttle transfer

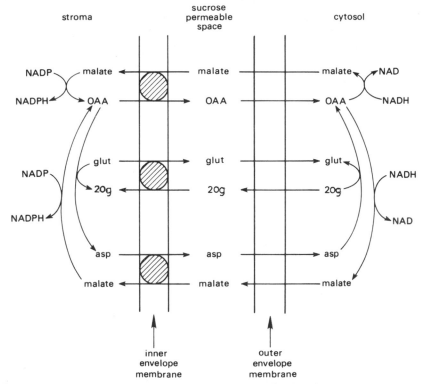

Figure 5.5.   Diagram of the dicarboxylate translocator showing the shuttle system for the transport of reducing power across the chloroplast envelope. Aspartate (asp); glutamate (glut); 2-oxoglutarate (2og).

of oxaloacetate and malate can provide a means of transporting reducing equivalents across the chloroplast envelope (Heber, 1974). This shuttle can operate in both directions depending on the redox potential of pyridine nucleotides outside and inside the chloroplast and involves the participation of both the chloroplast and cytoplasmic forms of malate dehydrogenase (Figure 5.5). Thus, for example, the import and reduction of oxaloacetate in the chloroplast concomitant with export and oxidation of malate in the cytoplasm constitutes a complete cycle, transferring one reducing equivalent from the chloroplast to the cytoplasm. The cycle may not operate in the same fashion in the dark since the chloroplast NADP–malate dehydrogenase is inactivated in darkness and NAD–malate dehydrogenase may function in its place (Heber, 1974). Aspartate and 2-oxoglutarate may be exchanged across the dicarboxylate translocator for glutamate and malate (see Section 5.9) and may provide a system of transfer for reducing equivalents (Figure 5.5), which is more efficient than the simple malate–oxaloacetate shuttle since the oxaloacetate concentration of the plant cell is very low. In C4 plants that form malate the rate of exchange of oxaloacetate for malate by the dicarboxylate translocator of the mesophyll chloroplasts is very similar to the rate of photosynthesis. Here the product of CO_2 fixation by phosphoenolpyruvate carboxylase in the cytoplasm is oxaloacetate. In the mesophyll chloroplast of the C4 plant it is reduced to malate, which is subsequently transported to the bundle sheath cells.

5.8. OTHER TRANSLOCATORS

It is clear that in mature chloroplasts of C3 plants the major movement of adenine nucleotides between the chloroplast and cytoplasm is not by direct transfer. Direct transfer of adenine nucleotides across the chloroplast envelope does occur but it is usually very slow. The average rate of transport of adenine nucleotides through the adenylate translocator is 5 μmole/h·mg chlorophyll. Transfer proceeds by counterexchange and is highly specific for ATP, the transport of ADP into the stroma being much slower than that of ATP. The main function of the carrier perhaps may be to supply the chloroplasts with ATP, for example, in the dark. It does not appear to be involved in the photophosphorylation of cytoplasmic ADP. Pyrophosphate is slowly transported in exchange for stromal adenylates. However, Robinson and Wiskich (1977) showed that the rate of ATP uptake by chloroplasts prepared from young pea shoots is in excess of 20 μmole/h·mg chlorophyll. Thus the activity of the adenylate carrier may be dependent on the developmental state of the leaves from which the chloroplasts are prepared. The inhibition of CO_2 fixation in isolated intact pea chloroplasts by pyrophosphate indicates that in young pea leaves the adenylate translocator is more active than in mature leaves. Huber and Edwards (1976) have also reported the presence of an ATP translocator in mesophyll chloroplasts by the C4 plant *Digitaria sanguinalis* that exhibits rates of ATP transport in the order of 40 μmole/h · mg chlorophyll.

There is a glucose carrier on the inner envelope membrane that appears to function in exporting D-glucose (but not L-glucose) produced by starch degradation in the dark from the chloroplast. Several hexoses and pentoses are transported across the envelope by this translocator. Glucose transport appears to be inhibited by a high external glucose concentration and therefore this sugar can be used as an osmoticum for chloroplast isolation media. The capacity of the glucose transport under optimum conditions has been estimated to be 60 μmole/h/mg chlorophyll.

Several other carriers have been observed and studied; these include transporters for glycine and serine (Nobel and Cheung, 1972), leucine and isoleucine (McLaren and Barber, 1977), pyruvate (Heber and Edwards, 1977), glycerate (Robinson, 1972), and ascorbate (Anderson et al., 1983).

5.9.   GRADIENT COUPLING

Heber (1974) and Heber and Heldt (1981) have stressed the importance of the presence of the gradients of glycollate and photosynthetic intermediates that exist between the chloroplasts, cytoplasm, and mitochondria. Gradients of metabolites across the chloroplast envelope are coupled to the proton gradient such that an opposite anion gradient is produced, for example, with the salts of weak acids where the anion is nonpenetrating and its protonation product is penetrating. Aliphatic acids such as acetate and the plant hormone, abscisic acid, rapidly permeate the chloroplast envelope via simple diffusion of their protonated ions. However, when isolated intact chloroplasts are illuminated in the presence of, for example, abscisic acid, transport and accumulation is enhanced. The pH gradient across the envelope in the light induces, at equilibrium, a rapid uptake of the undissociated form of the weak acid while the anionic form is relatively impermeable. Thus, the distribution of abscisic acid and other weak acids between the stroma and the medium, which can be predicted by the Henderson–Hasselbach equation, is inversely proportional to the distribution of protons. The extent of accumulation is dependent on the magnitude of the proton gradient. In the illuminated plant mesophyll cell the metabolite shuttle systems operate simultaneously and gradients of metabolites and ions are created between adjacent cellular compartments.

The subcellular distribution of total adenine nucleotides has been calculated by Stitt et al. (1982) to be approximately 47% in the chloroplast, 44% in the cytoplasm, and 9% in the mitochondria. During darkness the ATP required for cytoplasmic metabolism is supplied by mitochondrial oxidative phosphorylation. Mitochondrial electron transport generates an electrical potential difference across the inner mitochondrial membrane and this, in turn, drives oxidative phosphorylation and also ATP transport through the adenylate translocator against a concentration gradient. This allows a much higher ATP:ADP ratio to be produced in the cytoplasm than in the mitochondrial matrix. It is generally believed that in the illuminated leaf mesophyll cells the ATP in the cytoplasm is provided by export of triose phosphate from the chloroplasts. Rotenone, which inhibits mitochondrial electron transport, lowers the cytoplasmic ATP level in the dark but has no marked effect on cytoplasmic ATP in the light. Upon illumination the phosphorylation potential of the chloroplasts is increased by photophosphorylation. The phosphate translocator links the intra- and extrachloroplast adenylate systems (and pyridine nucleotide pools) through the shuttle systems involving dihydroxyacetone phosphate and PGA such that the export and subsequent oxidation of triose phosphate results in the coupled transfer of reducing equivalents and ATP into the cytoplasm. Whether mitochondrial oxidative phosphorylation also contributes to the cytoplasmic ATP pool in the light is still uncertain. In the past it has been considered that mitochondrial respiration, which is regulated by the phosphorylation potential, [ATP]:[ADP][Pi] or the [ATP]:[ADP] ratio, is inhibited during illumination of mesophyll cells. However, this assumption is probably incorrect (Goller et al., 1982). In the light ADP is rapidly converted into ATP in the chloroplast, and via the shuttle mechanisms, the demand for ADP is rapidly transduced to the cytosol. However, the activity of adenylate kinase (Birkenhead et al., 1982), in the intramitochondrial space, buffers the ADP demand of the chloroplasts and would allow ADP to be available to the mitochondria. Mitochondrial electron transport in the light, therefore, would not appear to be inhibited by a lack of ADP.

The pH and redox gradients that exist across the chloroplast envelope may maintain the phosphorylation potential of the cytoplasm at a higher value than that of the stroma. When the import of ATP into the cytoplasm exceeds that required for metabolism, the cytoplasmic phosphorylation potential will rise. During illumination the measured stromal ATP:ADP ratios are higher (ATP:ADP = 1–3) than in the dark (ATP:ADP = 0.2–1.0), but they are only approximately half the values that are attained in the cytoplasm. The phosphorylation potentials of the cytoplasm and mitochondria are linked through the mitochondrial adenylate carrier. Mitochondrial respiration is largely controlled by the phosphorylation potential and it is suggested that it is therefore suppressed by an increase in the cytoplasmic ATP:ADP ratio in the light via respiratory control by adenylates. Control of this nature would require a substantial decrease in the cytoplasmic ATP:ADP ratio in the dark to stimulate respiration. The extra-chloroplast ATP:ADP ratio has been reported to be higher in the light than in the dark (Hampp et al., 1982). Stitt et al. (1982) have questioned the validity of this data and have reported that the cytoplasmic ATP:ADP ratio does not necessarily decrease in the dark and may even increase. They suggest that when mitochondria and cytoplasm are measured together the ATP:ADP ratio in the dark is found to be lower than that in the light, but when the cytoplasm is measured alone then the ATP:ADP ratio need not change substantially.

Phosphorylation of NAD to NADP has been shown to occur in isolated chloroplasts during illumination (Ogren and Krogman, 1965; Muto et al., 1981). In spite of this the measured totals of the $NADP^+$–NADPH pool in the stroma have been found to remain relatively constant when leaves and isolated chloroplasts are illuminated NADP reduction in the stroma is rapid, however, after the initial lag phase of photosynthesis it is not uncommon for the [NADPH]:[NADP] ratio to decrease such that in steady-state photosynthesis this ratio is similar if not lower than that obtained in darkness (Takahama et al., 1981; Heber et al., 1982).

The phosphate translocator connects the pyridine nucleotide pool of that stroma with that of the cytosol. The cytosolic NAD pool is maintained in the reduced state by the action of a cytosolic nonreversible (nonphosphorylating) GAP dehydrogenase and glucose-6-phosphate dehydrogenase. Triose phosphate exported from the chloroplasts can be oxidized in the cytosol to yield PGA, ATP, and NADH (Figure 5.4). The ATP generated in the cytosol by this means is required for anabolic processes such as sucrose synthesis but the fate of the NADH has yet to be determined. Cytosolic NADH is oxidized by oxaloacetate forming malate, which can subsequently reenter the chloroplast in exchange for OAA. The malate–oxaloacetate and malate–aspartate shuttles may act as transfer systems by which reducing equivalents may return to the stroma. In physiological conditions the ratio of oxaloacetate to malate in equilibrium with the pyridine nucleotides may be so low (10^{-3}–10^{-4}) that the oxaloacetate transport by the dicarboxylate translocator would be severely restrained. The capacity of the malate–oxaloacetate shuttle alone is too low to mediate an effective shuttle of reducing equivalents between the chloroplast and the cytosol (Giersch, 1982). The dicarboxylate translocator is capable of transferring several dicarboxylate acids such as malate, oxaloacetate, 2-oxoglutarate, aspartate, and fumarate on an exchange basis (Lehner and Heldt, 1978). Aspartate aminotransferase, which is present in both the stroma and cytoplasm, catalyses the reaction of oxaloacetate and glutamate (Glu) to form aspartate (Asp) and 2-oxoglutarate (2OG). By this means the stromal NADP pool may be linked with the cytoplasmic NAD system and may be represented at equilibrium as

$$\underset{\text{cytoplasm}}{\frac{(\text{Asp})\,(\text{2OG})\,(\text{H}^+)\,(\text{NADH})}{(\text{Glu})\,(\text{Mal})\,(\text{NAD}^+)}} = \underset{\text{stroma}}{\frac{(\text{Asp})\,(\text{2OG})\,(\text{H}^+)\,[\text{NAD(P)H}]}{(\text{Glu})\,(\text{Mal})\,[\text{NAD(P)}]}} \qquad (5.8)$$

This shuttle system, however, has not been fully experimentally characterized and its transfer capacity remains to be accurately determined.

Isolated plant mitochondria will oxidize added NADH and it may be possible that during illumination cytosolic NADH may be used to produce ATP in the mitochondria. This depends on whether the cytosolic phosphorylation potential in the light is sufficient to suppress mitochondrial electron transport. Nitrate reduction in the cytosol (see Section 6.7) has been shown to be coupled to the transfer of reducing equivalent from the chloroplast via the shuttle systems (House and Anderson, 1980; Rathnam, 1978). Reducing equivalents for NO_3^- reduction in the light appear to be primarily derived from the chloroplast. The light-generated increase in the cytosolic NADH could suppress the competition between nitrate reduction and mitochondrial respiration for reducing equivalents that is significant in the dark and allow nitrate reduction to occur under aerobic conditions (Reed and Canvin, 1982).

REFERENCES

Anderson, J. M., Foyer, C. H., and Walker, D. A. (1983). *Planta* **158**, 442–450.

Apel, K. (1979). *Eur. J. Biochem.* **97**, 183–188.

Birkenhead, K., Walker, D. A. and Foyer, C. (1982). *Planta* **156**, 171–175.

Block, M. A., Joyard, J., and Douce, R. (1980). *Biochim. Biophys. Acta* **631**, 210–219.

Block, M. A., Dorne, A. J., Joyard, J., and Douce, R. (1983). *FEBS Lett.* **153**, 377–381.

Brown, E. G., Edwards, M. J., Newton, R. P., and Smith, C. J. (1980). *Phytochemistry* **19**, 23–30.

Cline, K., Andrews, J., Mersey, B., Newcomb, E. H., and Keegstra, K. (1981). *Proc. Natl. Acad. Sci. USA* **78**, 3595–3599.

Charlton, J. M., Treharne, K. J., and Goodwin, T. W. (1967). *Biochem. J.* **105**, 205–212.

Chua N.-H. and Schmidt, G. W. (1978). *Proc. Natl. Acad. Sci. USA* **75**, 6110–6114.

Costes, C., Burghoffer, C., Joyard, J., Block, M., and Douce, R. (1979). *FEBS Lett.* **103**, 17–21.

Dorne, A. J., Block, M. A., Joyard, J., and Douce, R. (1982a). *FEBS Lett.* **145**, 30–34.

Dorne, A. J., Block, M. A., Joyard, J., and Douce, R. (1982b). In *Biochemistry and Metabolism of Plant Lipids* (J. F. G. M. Wintermans and P. J. C. Kuiper, eds.), pp. 153–164. Elsevier/ North Holland Biomedical Press, New York.

Douce, R. (1974). *Science* **183**, 852–853.

Douce, R., Holtz, R. B., and Benson, A. A. (1973). *J. Biol. Chem.* **248**, 7215–7222.

Douce, R. and Joyard, J. (1979). *Adv. Bot. Res.* **7**, 1–116.

Douce, R. and Joyard, J. (1981). In *Photosynthesis* (G. Akoyunoglou, ed.), Vol. III. Structure and Molecular Organization of the Photosynthetic Membrane, pp. 187–198. Balaban International Science Services, Philadelphia.

Douce, R. and Joyard, J. (1982). In *Methods in Chloroplast Molecular Biology* (M. Edelman, R. Hallick, and N.-H. Chua, eds.), pp. 239–256. Elsevier/North Holland Biomedical Press, Amsterdam, New York.

Douce, R. and Joyard, J. (1983). In *Chloroplast Biogenesis* (N. Baker and J. Baker, eds.), Topics in Photosynthesis, Vol. 5. Elsevier/North Holland Biomedical Press, New York.

Flugge, U. I. (1982). *FEBS Lett.* **140**, 273–276.

Flugge, U. I. and Heldt, H. W. (1977). *FEBS Lett.* **82**, 29–33.

Flugge, U. I. and Heldt, H. W. (1978). *Blochem. Biophys. Res. Commun.* **84**, 37–44.

Flugge, U. I. and Heldt, H. W. (1981). *Biochim. Biophys. Acta* **638**, 296–306.

Giersch, C. (1982). *Arch. Biochem. Biophys.* **219**, 379–387.

Goller, M., Hamp, R., and Ziegler, H. (1982). *Planta* **156**, 255–263.

Hampp, R., Goller, M., and Ziegler, H. (1982). *Plant Physiol.* **69**, 448–455.

Heber, U. (1974). *Ann. Rev. Plant Physiol.* **25**, 393–421.

Heber, U., Kirk, M. R., Gimmler, H., and Schafer, G. (1974). *Planta* **120**, 31–46.

Heber, U. and Heldt, H. W. (1981). *Ann. Rev. Plant Physiol.* **32**, 139–168.

Heber, U. and Walker, D. A. (1979). *Trend. Biochem. Sci.* **4**, 252–256.

Heber, U., Takahama, U., Neimanis, S., Shimizu-Takahama, M. (1982). *Biochim. Biophys. Acta* **679**, 287–299.

Heldt, H. W. and Rapley, L. (1970). *FEBS Lett.* **10**, 143–148.

Highfield, P. E. and Ellis, R. J. (1978). *Nature* **271**, 420–424.

House, C. M. and Anderson, J. W. (1980). *Phytochemistry* **19**, 1925–1930.

Huber, S. C. (1978). *Plant Physiol.* **62**, 321–325.

Huber, S. C. and Edwards, G. E. (1976). *Biochim. Biophys. Acta* **440**, 675–687.

Huber, S. C. and Edwards, G. E. (1977). *Biochim. Biophys. Acta* **462**, 603–612.

Joyard, J. and Douce, R. (1977). *Biochim. Biophys. Acta* **486**, 273–285.

Joyard, J., Grossman, A. R., Bartlett, S. G., Douce, R., and Chua, N.-H. (1982). *J. Biol. Chem.* **257**, 1095–1101.

Kaiser, W. M., Urbach, W., and Gimmler, H. (1980). *Planta* **143**, 170–175.

Leech, R. M. and Murphy, D. J. (1976). In *The Intact Chloroplast* (J. Barber, ed.), pp. 365–401. Elsevier/North Holland Biomedical Press, New York.

Lehner, K. and Heldt, H. W. (1978). *Biochim. Biophys. Acta* **501**, 531–544.

Lichtenthaler, H. A., Prenzal, U., Douce, R., and Joyard, J. (1981). *Biochim. Biophys. Acta* **641**, 99–105.

Mackender, R. O. and Leech, R. M. (1970). *Nature* **228**, 1347–1349.

McLaren, J. S. and Barber, D. J. (1977). *Planta* **136**, 147–151.

Murakami, S. and Strotman, H. (1978). *Arch. Biochem. Biophys.* **185**, 30–38.

Muto, S., Miyachi, S., Usuda, H., Edwards, G. E., and Bassham, J. A. (1981). *Plant Physiol.* **68**, 324–328.

Nobel, P. S. and Cheung, Y. S. (1972). *Nature* **237**, 207–208.

Ogren, W. L. and Krogman, D. W. (1965). *J. Biol. Chem.* **240**, 4603–4608.

Porter, J. W. and Spurgeon, S. L. (1979). *Pure Appl. Chem.* **51**, 609–622.

Rathnam, C. K. M. (1978). *Plant Physiol.* **62**, 220–223.

Reed, A. J. and Canvin, D. T. (1982). *Plant Physiol.* **69**, 508–513.

Robinson, S. P. (1982). *Plant Physiol.* **70**, 1032–1038.

Robinson, S. P. and Wiskich, J. T. (1977). *Plant Physiol.* **59**, 422–427.

Sanchez, J. and Mancha, M. (1981). *Planta* **153**, 519–523.

Soll, J., Douce, R., and Schultz, G. (1980). *FEBS Lett.* **112**, 243–246.

Sprey, B. and Laetsch, W. M. (1976). *Z. Pflanzenphysiol.* **78**, 146–163.

Steiger, H. M., Beck, E., and Beck, R. (1977). *Plant Physiol.* **60**, 903–906.

Stitt, M., Lilley, R. McC., and Heldt, H. W. (1982). *Plant Physiol.* **70**, 971–977.

Takahama, U., Shimizu-Takahama, M., and Heber, U. (1981). *Biochim. Biophys. Acta* **637**, 530–539.

Tobin, E. (1981). *Plant Mol. Biol.* **1**, 35–51.

Van Besouw, A. and Wintermans, J. F. G. M. (1978). *Biochim. Biophys. Acta* **529**, 44–53.

6

THE END PRODUCTS OF PHOTOSYNTHESIS IN LEAVES

6.1. STARCH AND SUCROSE

Under saturating light the major end products of CO_2 fixation in leaves and leaf protoplasts are starch and sucrose. Most of the photosynthetically fixed carbon is exported to the cytoplasm in the form of sugar phosphates to be either utilized in metabolism or converted into a transportable form. Sucrose occupies a central position in plant metabolism (Pontis, 1977). Only a small number of photosynthetic eukaryotes, species of the red and brown algae, for example, do not synthesize sucrose. In general, triose phosphates are the major end products of photosynthesis in the chloroplasts and sucrose is the major photosynthetic product of leaves. Sucrose is synthesized to transport assimilate from the source tissues to sink tissues throughout the plant to serve as a source of organic carbon or, in some cases, to be accumulated and stored (Avigad, 1982). However, during periods of high net CO_2 fixation production of triose phosphate by the chloroplasts exceeds export and this excess is temporarily converted into starch. Starch is the only product of photosynthesis that is retained by the chloroplast. The starch reserve of the chloroplast is mobilized and utilized by the plant in darkness and at times of limited photosynthesis. Sucrose exported from the leaves is usually the major carbon source for starch synthesized in reserve tissues.

Sucrose is an α-glucopyranosly-β-D-fructofuranoside (Figure 6.1). It is a nonreducing disaccharide that can be hydrolyzed to glucose and fructose by dilute acids or by the action of the enzyme invertase. The β-fructofuranoside nature of sucrose confers a very high free energy of hydrolysis on the linkage between the two sugars. The free energy of hydrolysis of sucrose ($\Delta G^{\circ\prime}$ of -7.0 kcal/mole) is close to that of the γ-phosphoryl group of ATP. Most other glycosides have much lower $\Delta G^{\circ\prime}$ values.

Sucrose is a highly soluble sugar that can be accumulated to considerable concentrations without an apparent inhibitory effect on most of the biochemical reactions of plant metabolism. The sucrose molecule is electrically neutral and therefore does not readily interact electrostatically with other charged molecules and as a nonreducing sugar it is free from interactions with functional groups of other biological molecules. These properties make sucrose a useful component of the mechanisms that serve to regulate osmotic pressure and water relationships between the cellular compartments in the plant. Sucrose is also readily transported across biological membranes (Komer, 1982). Arnold (1968) has suggested that sucrose provides a convenient and comparatively unreactive, and therefore protected, derivative of glucose.

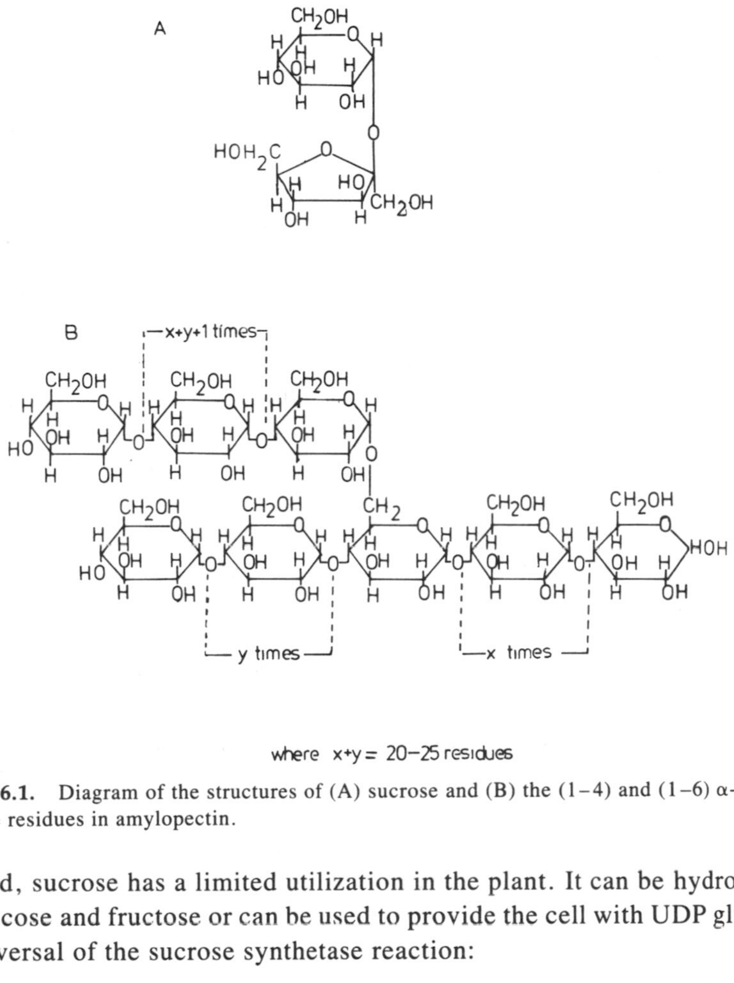

Figure 6.1. Diagram of the structures of (A) sucrose and (B) the (1–4) and (1–6) α-linked glucose residues in amylopectin.

Indeed, sucrose has a limited utilization in the plant. It can be hydrolyzed to glucose and fructose or can be used to provide the cell with UDP glucose by reversal of the sucrose synthetase reaction:

$$\text{UDP glucose} + \text{D-fructose} \rightleftharpoons \text{sucrose} + \text{UDP} + \text{H}^+ \quad (6.1)$$

This reaction retains the energy of the glycoside bond in the α-glucosyl of the UDP glucose molecule. In some plants sucrose can serve as the fructosyl donor as well as the primary acceptor for the synthesis of oligosaccharides of the inulin-type of the acceptor in the synthesis of oligosacchardes of the raffinose series (Kandler and Hopf, 1982).

Sucrosyl oligosaccharides function as storage materials and also appear to have a cryoprotective function contributing to frost resistance in some plant species. Frost resistant plants show a decrease in starch synthesis and an increase in the production of low molecular weight oligosaccharides at the onset of winter. Sucrose, raffinose, and stachyose are often accumulated during such cold acclimatization.

Higher plants translocate oligosaccharides and/or sugar alcohols. Sucrose predominates in sugar transport but there are exceptions, for example, sorbitol is the primary and major photosynthetic product transported by woody Rosaceae species and raffinose-type oligosaccharides are found to be translocated in some plants. From the site of synthesis assimilate crosses to the phloem sieve tubes for transport away from the leaves. Transfer of sucrose to the sieve tubes have been shown to be not entirely symplastic. Sucrose can be actively loaded from the apoplast into the companion cell sieve tube complexes of minor veins. The majority of photosynthetic cells have the capacity to transport sucrose across the plasmalemma but it is considered that *in vivo* sucrose is released from the cytoplasm into the free space surrounding the photosynthetic cells only in the immediate vicinity of the phloem (Huber and Moreland, 1980). In some C3 species there are laterally orientated paraveinal mesophyll cells at the level of the veins within the spongy mesophyll and these may be important sites of transfer of assimilate to the phloem. All C3 and C4 plants have a functional bundle sheath that must be involved in the transfer of sucrose to the phloem. The export of sucrose from the photosynthetic cells into the cell space is regulated and may be achieved by a sugar-K^+ symport mechanism that has an energy requirement (Huber and Moreland, 1981).

Sucrose does not accumulate in the free space surrounding the cells in the leaf but concentrates in adjacent sieve elements where it reaches concentrations of up to $1\,M$. Selective phloem loading from the apoplast involving cotransport of sucrose and protons has been demonstrated (Giaquinta, 1977a; Oelrot and Bonnemain, 1981). This involves an electrochemical potential gradient of protons established by a membrane ATPase (Giaquinta 1977b; 1979), which acts as a proton pump (Komer, 1982). Sucrose moves rapidly through the phloem (0.5–1.0 m/h) and is unloaded into cells of competing sink tissue.

Starch, which generally consists of a mixture of amylose and amylopectin, is a common constituent in almost all higher plant organs. The reserve carbohydrates of many algae are polymers of α-D-glucose, which structurally resemble higher plant starches and amylopectins (Manners and Sturgeon, 1982). The extent of starch accumulation in leaves of higher plants varies considerably among plant species. In turn starch accumulation within the chloroplasts of given species is influenced by the rate of assimilate trans-location from the leaf. Amylose, which accounts for 20–25% of most starches, consists of long chains of (1–4) α-linked D-glucose residues and can range in size from several hundred to several thousand residues (Figure 6.1) depending on the plant species. The major component of many starches is amylopectin, which has a branched structure. Linear chains of (1–4) linked α-D-glucose residues approximately 20–25 residues in length are interlinked by (1–6) α-D-glucosidic interchain linkages. Amylopectin can form very large polymers with molecular weights of up to 10^8 in which thousands of chains are interlinked together.

6.2. THE BIOSYNTHESIS OF SUCROSE

Sucrose is synthesized *de novo* in the leaf cytoplasm from triose phosphate (Figure 6.2) exported from the chloroplast in exchange for orthophosphate. The triose phosphates dihydroxyacetone phosphate and glyceraldehyde phosphate firstly undergo aldol condensation to FBP in a freely reversible reaction catalyzed by FBP aldolase (EC. 4.1.2.13). FBP is subsequently hydrolyzed to F6P by the action of cytoplasmic FBPase (EC 3.1.3.11), which is clearly distinct from the stromal isoenzyme. Phosphoglucose isomerase (EC 5.3.1.9) converts F6P to its isomer glucose-6-phosphate from which glucose-1-phosphate can be produced by the action of phosphoglucomutase (EC 2.7.5.1). These last enzymes catalyze freely reversible reactions and at equilibrium the hexose phosphates are maintained in a ratio of 10:20:1 molecules of F6P:glucose-6-phosphate:glucose-1-phosphate. UDP–glucose pyrophosphorylase (EC 2.7.7.9) acts as a sink for glucose-1-phosphate, which pulls carbon flow toward sucrose formation. Glucose-6-phosphate occupies a central position in several metabolic pathways that provide the cell with NADPH, ATP, and precursors. It can be oxidized through the oxidative pentose phosphate pathway providing the cell with NADPH, or metabolized to CO_2 and H_2O yielding ATP, or may form the precursor for polysaccharide biosynthesis. The enzyme UDP–glucose pyrophosphorylase catalyzes the freely reversible transfer of a uridyl moiety between specific phosphate acceptors:

$$\text{UDP glucose} + \text{pyrophosphate} \rightleftharpoons \text{UTP} + \text{glucose-1-phosphate}$$

$$(6.2)$$

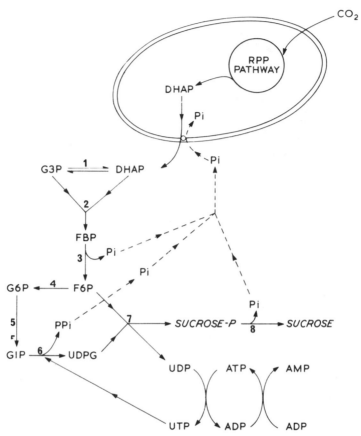

Figure 6.2.   The sucrose synthetic pathway. The enzymes involved are (1) triosephosphate isomerase, (2) fructosebisphosphate aldolase, (3) FBPase, (4) phosphoglucose isomerase, (5) phosphoglucomutase, (6) UDP–glucose pyrophosphorylase, (7) sucrose–phosphate synthetase, and (8) sucrose–phosphate phosphatase.

However, this reaction is coupled to that of highly active pyrophosphatases and constitutes an essentially irreversible step towards sucrose synthesis. UDP–glucose pyrophosphorylase often constitutes the major phosphorylase activity in plants, its activity being generally 30- to 50-fold higher than that of ADP–glucose pyrophosphorylase. The plant enzymes studied so far show a high degree of specificity toward UDP glucose, UTP, and glucose-1-phosphate as substrates. The enzyme shows an absolute requirement for divalent cations, Mg^{2+} giving optimal activity at concentrations of $1-2$ mM.

The enzymes directly involved with sucrose biosynthesis in the leaf are sucrose phosphate synthetase and sucrose phosphate phosphatase. Sucrose phosphate synthetase was first described by Leloir and Cardini (1955) and catalyzes the synthesis of sucrose phosphate:

$$\text{UDP glucose} + \text{F6P} \rightleftharpoons \text{sucrose-phosphate} + \text{UDP} \quad (6.3)$$

The sucrose phosphate produced by this reaction is rapidly cleaved in the cytoplasm by a specific sucrose phosphate phosphatase (Hawker and Hatch, 1966) to yield free sucrose:

$$\text{Sucrose phosphate} \longrightarrow \text{sucrose} + \text{Pi} \quad (6.4)$$

Sucrose phosphate phosphatase is present in the cytoplasm of the leaf at considerably higher concentrations than sucrose phosphate synthetase (Pollock, 1976). This ensures that the sucrose–phosphate concentration in the cytoplasm is maintained at a very low level.

It is most likely that sucrose phosphate synthetase and the cytoplasmic FBPase play important roles in the regulation of carbon flow to sucrose. Sucrose phosphate synthetase displays several properties that suggest how its activity may be regulated. The enzyme from spinach leaves has an ordered sequential reaction mechanism (Harbron et al., 1981). In the light the steady-state concentration of F6P is approximately 5–10 mM and therefore this substrate is present at saturating concentrations during illumination. Amir and Preiss (1982) have found that the F6P saturation curve for the spinach leaf enzyme is sigmoidal in nature, indicating multiple and interacting sites for F6P. However, other workers have found that the substrate saturation curve for spinach leaf sucrose–phosphate synthetase is hyperbolic with respect to F6P.

The concentration of UDP glucose in spinach leaf cytoplasm has been estimated to be approximately 2 mM and at best is barely saturating for sucrose–phosphate synthetase. The poor affinity of the enzyme for UDP glucose could have a profound effect on the activity of sucrose–phosphate synthetase. The rate of sucrose–phosphate production may be directly related to the cytosolic concentration of UDP glucose. In these circumstances the rate of the reaction may be fundamentally dependent on the supply of F6P and ATP, for recycling UDP to produce UDP glucose. UDP is a potent inhibitor of spinach leaf sucrose–phosphate synthetase with K_i of approximately 0.7 mM. Consequently, failure to recycle UDP will produce a dual inhibitory action on the enzyme firstly, by direct inhibition and secondly by the reduction of the UDP–glucose concentration. Pi is a competitive inhibitor of sucrose–phosphate synthetase activity with respect to UDP glucose ($K_i \approx 10$ mM). FBP is also a inhibitor of spinach leaf sucrose–phosphate synthetase with a K_i of approximately 0.8 mM. UDP and Pi could therefore provide mechanisms for the fine control of sucrose–phosphate synthetase activity.

Mature leaves are often found to contain a low sucrose synthetase activity (EC 2.4.1.13). This enzyme catalyzes a freely reversible reaction and is considered to be involved primarily in the synthesis of UDP glucose. The role of this enzyme in leaves has yet to be demonstrated. Recently a UDPase enzyme that catalyses reaction (6.5) has been observed in soybean leaves (Huber and Pharr, 1981):

$$\text{UDP} \longrightarrow \text{UMP} + \text{Pi} \qquad (6.5)$$

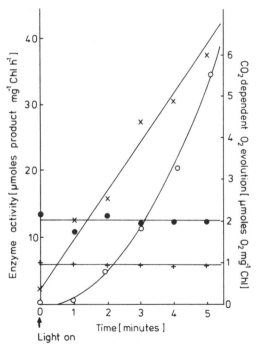

Figure 6.3.   The effect of light on sucrose phosphate synthetase and cytoplasmic FBPase in wheat protoplasts. CO_2-dependent O_2 evolution (○), sucrose–phosphate synthetase (+), chloroplast FBPase (×), and cytoplasmic FBPase (●). (Taken from Foyer, C., Walker, D., and Latzko, E. (1982). *Z. Pflanzenphysiol.* **107**, 457–465.)

This enzyme has been postulated to have a function in sucrose biosynthesis since it will reduce the concentration of the inhibitor, UDP.

Control of carbon flow to sucrose is also regulated by modification of the cytoplasmic FBPase, which catalyzes

$$FBP + H_2O \longrightarrow F6P + Pi \qquad (6.6)$$

This enzyme requires free Mg^{2+} or Mn^{2+} for activity and is subject to AMP inhibition, which can be relieved by high Mg^{2+} concentrations. The enzyme has a high affinity for its substrate ($K_m = 3\ \mu M$, Zimmermann et al., 1978) but at concentrations above 100 μM FBP is inhibitory to enzyme activity. F6P at concentrations above 2 mM and millimolar concentrations of orthophosphate also result in an inhibition of the enzyme activity. Neither the cytoplasmic FBPase nor sucrose phosphate synthetase are light activated; the maximum activity of both enzymes is constant in light and darkness (Figure 6.3). However, fluctuations in the cytosolic concentration of fructose-2,6-biphosphate, which is a competitive inhibitor of the cytosolic FBPase, are most probably involved in regulation of the enzyme activity during darkness and illumination (Weeden et al., 1982). Fructose-2,6-bisphosphate also enhances the inhibition produced by AMP (Stitt et al., 1982).

6.3. STARCH SYNTHESIS

During photosynthesis some of the photosynthetic product is retained temporarily in the chloroplast as starch. Although starch normally accumulates in the light and is depleted in the dark, synthesis and degradation have been shown to occur simultaneously (Stitt and Heldt, 1981). The inhibition of starch synthesis in the dark is reversible since starch will accumulate in the chloroplasts in the dark when leaf disks are supplied with exogenous sugars (Herold and Walker, 1979). It is generally accepted that under conditions where CO_2 fixation exceeds the rate at which triose phosphate can be converted into sucrose in the cytosol, starch synthesis in the chloroplasts is increased. In isolated chloroplasts starch accumulation is enhanced when export of triose phosphate is restricted. The amount of starch present in the chloroplast may itself have an influence on the rates of starch synthesis and breakdown since starch accumulation often declines or can even cease toward the end of a period of illumination.

Starch is synthesized in the stroma from triose phosphate produced by CO_2 fixation. The initial reactions of starch biosynthesis are the same as those involved in sucrose synthesis in the cytoplasm. Glucose-1-phosphate is formed by a pathway similar to that occurring in sucrose synthesis (Figure 6.2) and is converted into ADP glucose by the enzyme ADP–glucose pyrophosphorylase, which catalyzes

$$\text{glucose phosphate} + \text{ATP} \longrightarrow \text{ADP glucose} + \text{pyrophosphate}$$

$$(6.7)$$

ADP–glucose pyrophosphorylases from leaves and several bacteria have been found to be tetrameric proteins of approximately 210,000 molecular weight. The enzymes require Mg^{2+} for activity and appear to be highly regulated. In leaves ADP–glucose pyrophosphorylase is activated by the primary product of CO_2 fixation, PGA (Ghosh and Preiss, 1965). PGA increases the apparent affinity of the spinach leaf enzyme for its substrates although the affinity for Mg^{2+} is not changed. The substrate saturation curves of the spinach leaf enzyme are hyperbolic in the presence of PGA. The activation kinetics of the enzyme in the presence of PGA are hyperbolic in nature from pH 7.0–7.5 but become progressively sigmoidal as the pH increases to 8.5 (Preiss, 1982). Activation by PGA is common to all leaf ADP–glucose pyrophosphorylases so far studied. Phosphoenolpyruvate, F6P, and FBP are also activators but they are not as effective as phosphoglycerate. 20 μM PGA will produce 50% maximal activation of the purified spinach enzyme. During illumination the concentration of phosphoglycerate in the chloroplast has been found to be approximately 4.0 mM. The orthophosphate concentration of the stroma in the dark has been estimated to be between 5 and 10 mM although it is probably much lower in the light. Orthophosphate (Pi) is a potent inhibitor of leaf ADP–glucose pyrophosphorylases. The purified spinach leaf enzyme is inhibited by 50% by 22 μM Pi in the absence of activator at pH 7.5 (Ghosh and Preiss, 1966). PGA relieves this inhibition such that in the presence of 1 mM PGA, 1.3 mM Pi is required to produce 50% inhibition. Thus, the presence of the activator decreases the sensitivity of the enzyme to the inhibitor. The activation kinetics with PGA become sigmoidal in the presence of Pi and similarly the sigmoidicity of the Pi inhibition curve is enhanced by the presence of PGA.

The interaction between the inhibitor Pi and the activator PGA has been suggested to produce effective regulation of the ADP–glucose pyrophosphorylase such that *in vivo* the enzyme may be subject to regulation by the ratios of [PGA]:[Pi] in the stroma rather than the absolute concentration of either. This type of regulation of ADP–glucose pyrophosphorylase has been observed in starch synthesis in green algae and higher plants and also during glycogen synthesis in cyanobacteria (Preiss, 1982). ADP–glucose pyrophosphorylase is inhibited at low ATP:ADP ratios (Kaiser and Bassham, 1979) but the ATP:ADP ratio does not always change significantly between light and darkness.

Recondo and Leloir (1961) showed that the synthesis of new (1–4) α-glucosidic linkages of starch occurred predominantly by the transfer of the glucosyl moiety of ADP glucose to a preexisting primer such as maltose or other such oligosaccharides by the action of the enzyme starch synthase (Figure 6.3). The soluble starch synthases of reserve tissues and leaves are highly specific for ADP glucose. Starch synthases of reserve tissues and leaves are highly specific for ADP glucose. Starch synthases bound to starch granules in storage tissues will utilize other glucosyl donors such as UDP glucose but the activities are much less than the ADP glucose. The affinity of these enzymes for ADP glucose is up to 30-fold higher than that for UDP glucose suggesting that ADP glucose is the preferred substrate even for these particulate synthetases. The particulate starch synthase of leaves are specific for ADP glucose. A branching or Q enzyme forms (1–6) α-linkages probably by interchain transfer. Multiple forms of both starch synthases and branching enzymes (Hawker et al., 1974) appear to occur in plants although the function of the various forms is not known. The molecular weights of the spinach leaf branching enzymes varies between 80,000 and 89,000. Borovsky et al. (1976) proposed that the branching reaction firstly requires the formation of a double helix by the two amylose chains undergoing hydrolysis with subsequent transfer.

6.4. STARCH BREAKDOWN

Starch accumulated in the light is broken down in subsequent darkness. Starch breakdown occurs in the chloroplast by both phosphorylytic and hydrolytic pathways. In the hydrolytic sequence starch is degraded to free sugars that are released from the chloroplasts. α-amylase hydrolyses (1–4) α-glusocyl linkages to produce soluble dextrins, oligosaccharidies, and glucose. The dextrins and oligosaccharides can then be further hydrolyzed to glucose. The R-enzyme breaks (1–6) α-glucosyl linkages. The D-enzyme transfers glucosyl units between short chain dextrins catalyzing the reversible condensation of donor and acceptor (1–4) α-glucans releasing free glucose.

Starch phosphorylase catalyses the phosphorolysis of $(1-4)$ α-glucosyl chains to yield glucose-1-phosphate [reaction (6.8)], which can then be converted to triose phosphate.

$$(1-4) \ \alpha\text{-(glucosyl)}_{n+1} + \text{Pi} \longrightarrow (1-4) \ \alpha\text{-(glucosyl)}_n + \text{glucose-1-phosphate}$$
$$(6.8)$$

The pathways of starch mobilization are influenced by the Pi concentration in the medium surrounding the chloroplasts (Stitt and Heldt, 1981). When orthophosphate was omitted from the suspending medium of isolated intact spinach chloroplasts in the dark 80% of the starch degradation produced free sugars. The overall rate of starch breakdown was not significantly changed but low Pi appeared to switch chloroplast starch breakdown from phosphorolysis to hydrolysis. Phosphorolysis could be increased by the addition of Pi and may then account for up to half of the starch degraded.

6.5. PARTITIONING OF ASSIMILATE BETWEEN STARCH AND SUCROSE

A sink tissue is one that has a net import of assimilate; growing, storing, or actively metabolizing tissues can therefore act as sinks for sucrose. There is indirect evidence that suggests that the sink tissue is able to "communicate" its requirements to the source. However, the nature of the signal(s) is unknown. Since the phloem forms a link between the sink and source, changes in the activity of the sink could lead to pressure/flow changes in the gradients of sucrose in the sieve tubes. Loading of sucrose into the phloem may then be changed and could eventually have an influence on the rate of CO_2 fixation and assimilate partitioning between starch and sucrose. The extent of sucrose synthesis appears to be regulated to some degree by source/sink interactions. Daytime filling of the vacuole with sucrose can be observed. This store of sucrose is released in the following night. The vacuole provides a diurnal store of fixed carbon similar in size to that of the chloroplast starch pool and possibly fulfilling a similar function. The rate of translocation of assimilates from the leaf that is at least partially controlled by sink demand (Huber and Israel, 1982) can influence starch accumulation. For example, conditions such as nutrient deficiency cause starch accumulation whereas increased demand for assimilates such as that imposed by root nodulation cause greater partitioning of carbon into sucrose in the leaf cells. There could also be a genetic variation between species that have different capacities for sucrose biosynthesis. Sucrose–phosphate synthetase appears to be a key control point in this variation, since the amounts of this enzyme present in leaves are only sufficient to account for rates of sucrose formation. Leaf starch levels have been found to be negatively correlated with the cytoplasmic levels of sucrose–phosphate synthetase in leaves (Huber, 1981). Thus, a higher activity of sucrose–phosphate synthetase could result in higher rates of sucrose formation that, in turn, could promote rapid export of triose phosphate from the stroma and result in reduced starch accumulation. Similarly, lower activities of sucrose–phosphate synthetase would result in lower sucrose biosynthesis and diversion of carbon into starch. The sensitivity of sucrose–phosphate synthetase to inhibition by sucrose also appears to vary between species. Species that tend to accumulate starch have been found to have sucrose–phosphate synthetase which are sensitive to inhibition by sucrose (Huber, 1981), whereas the enzyme from other species shows little or no sensitivity to the presence of sucrose (Foyer et al., 1983).

Partitioning of photosynthetically fixed carbon between starch and sucrose in the mesophyll cell of the leaf is biochemically controlled (Herold, 1980). Chloroplast metabolism is intimately associated with the rate of sucrose biosynthesis since the reactions of the sucrose pathway liberate orthophosphate. The chloroplasts must reimport this orthophosphate in order to export triose phosphate via the phosphate translocator. Changes in the orthophosphate status of the cytoplasm have been shown to have a profound influence on the rate of photosynthesis and on starch formation. In certain leaves sequestration of Pi occurs by feeding mannose, a monosaccharide that occurs naturally only in small amounts (Herold and Lewis, 1977) and is phosphorylated in the cytoplasm by hexokinase to mannose phosphate, which cannot be metabolized further. Mannose feeding has an inhibitory action on photosynthesis and increases the starch content of the chloroplasts (Herold et al., 1976). This occurs because cytoplasmic Pi is reduced by mannose treatment and results in high stromal ratio of [PGA]:[Pi], which stimulates ADP–glucose pyrophosphorylase activity that in turn enhances starch biosynthesis. The export of triose phosphate from the chloroplast is decreased because Pi is not available for exchange across the phosphate translocator and therefore sugar phosphates are retained in the chloroplast and are available for starch synthesis (Figure 6.4). Simultaneously, the phosphorylitic pathway of starch breakdown is inhibited by the lower stromal Pi level. Changes in the cytosolic level of the regulator compound fructose-2,6-bisphosphate,

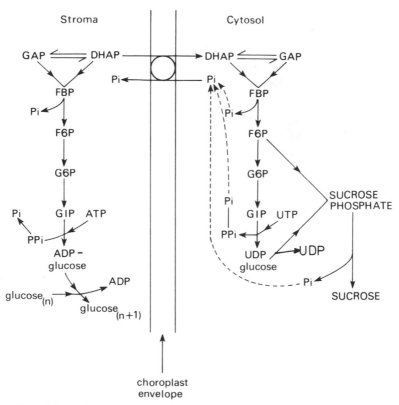

Figure 6.4. Diagram to show the central role of the phosphate translocator in determining the partitioning of carbon between starch synthesis in the stroma and sucrose synthesis in the cytosol.

which fluctuates during the photoperiod and upon light/dark transitions, are also involved in the regulation of assimilate partitioning. Micromolar concentrations of fructose-2,6-bisphosphate inhibit the cytoplasmic FBPase and will decrease carbon flow to sucrose.

In C4 plants the enzymes of the C4 pathway and the RPP pathway show a rigid compartmentation and it is probable that some of the enzymes of the carbohydrate and end-product pathways also have an asymmetric distribution between the bundle sheath and mesophyll cells in the leaves of such plants. In maize, sucrose–phosphate synthetase and cytoplasmic FBPase appear to be largely compartmented in the mesophyll cells. It is well-known that starch accumulates in bundle sheath chloroplasts during illumination

(Downton and Hawker, 1973). Starch accumulates in the mesophyll chloroplasts only after prolonged illumination.

6.6. FATTY ACID SYNTHESIS

The chloroplast is the major site of fatty acid synthesis in the leaf. The principal products of this process in the chloroplast are palmitic and oleic acids. These may be transported into the cytosol where they are further metabolized to linoleic and linolenic acids. They are then reimported into the chloroplasts where they constitute the principal membrane lipids. The mature plant synthesizes all its fatty acids as products of photosynthesis either directly in the leaves or in other plant organs from transported sugars. Once formed these fatty acids are not translocated to other parts of the plant. All the nonphotosynthetic tissues of the plant that have sites of fatty acid synthesis, for example, the proplastid of the seed, utilize photosynthetic product in the form of sucrose imported from the leaf. The principal fatty acid synthesized by the proplastids of lipid storing seeds such as castor bean is oleic acid.

The reactions of fatty acid synthesis in the chloroplast are essentially similar to those that occur in all living organisms (Stumpf, 1981). In the chloroplast acetyl coenzyme A is a precursor for the biosynthesis of fatty acids, β-carotene, prenyllipids, and plastoquinone-9. Chloropast acetyl coA can be produced either from imported acetate (Figure 6.5) or through glycolysis. Phosphopyruvate hydratase and pyruvate kinase have been localized in the chloroplasts (Stitt and ap Rees, 1979). A partial glycolytic pathway, at least, may therefore operate in chloroplasts although two enzymes of the pathway, phosphoglyceromutase and pyruvate dehydrogenase, appear to be absent from chloroplasts of some plants (Stitt and ap Rees, 1979; Murphy and Stumpf, 1981). Thomas et al. (1981) have suggested that carnitine is involved in the transport of acyl groups across membrane. Palmitoyl acyl-CoA synthetase and carnitine palmitoyltransferase have been localized in barley leaf etio-chloroplasts (Thomas et al., 1982). Long-chain acylcarnitine, formed by the action of these enzymes, may move more easily through membranes than the long chain acyl-CoA compound and thus facilitate the movement of acyl groups from one organelle to another in the leaf cell. An alternative pathway of acetyl-CoA synthesis in the chloroplast involving the import of free acetate released from the mitochondria appears to make

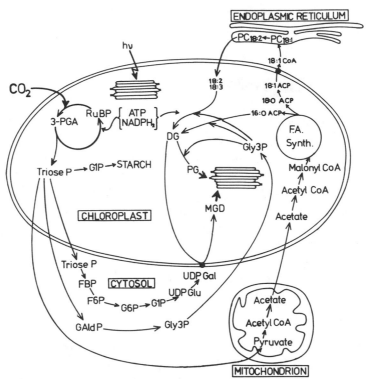

Figure 6.5. Diagram of a pathway of acyl lipid precursor biosynthesis in the chloroplast that involves collaboration among chloroplast, cytosolic, and mitochondrial enzymes. Acyl carrier protein (ACP); diacylglycerol (DG); fatty acid (FA); fructose bisphosphate (FBP); fructose 6-phosphate (F6P); glyceraldehyde-3-phosphate (GAldP); Glucose-6-phosphate (GlP); glycerol-3-phosphate (Gly3P); monogalactosyldiacylglycerol (MGD); phosphatidylglycerol (PG); 3-phosphoglyceric acid (PGA); ribulose bisphosphate (RuBP); UDP galactose (UDP-Gal); UDP-glucose (UDP-Glu); palmitic acid (16:0); stearic acid (18:0); oleic acid (18:1); linoleic acid (18:2); α linolenic acid (18:3). (Taken from Murphy, D. J. and Walker, D. A. (1982). *Planta* **156**, 84–88.)

a most important contribution to acetyl-CoA synthesis in chloroplasts. Free acetate readily diffuses across the chloroplast envelope. In the stroma it is esterified to CoA by a highly active acetyl-CoA synthetase, which has been shown to be present in the chloroplast (Kuhn et al., 1981).

The chloroplast envelope is relatively impermeable to CoA. Upon illumination isolated intact chloroplasts will synthesize acetyl-CoA *in vitro* from endogenous CoA when supplied with either ^{14}C-bicarbonate or ^{14}C-acetate and incorporate this radioactivity into fatty acids. Murphy and Walker

(1982) have suggested that free acetate imported from the mitochondria is the principal substrate for chloroplast acetyl-CoA biosynthesis in spinach, since acetate was found to be incorporated at a rate 40-fold in excess of that of bicarbonate when these substrates were supplied to illuminated intact chloroplasts at physiological concentrations. In common with all systems of fatty acid synthesis from acetyl CoA and malonyl CoA, the low molecular weight ACP facilitates the orderly addition of malonyl ACP to a lengthening acyl chain. The acyl moiety is covalently bonded to ACP via a thioester linkage; the dual reactivity of the thioester compound allows the reaction to proceed.

The systems involved in fatty acid synthesis in plants can be summarized by equations (6.9) and (6.10) (see also Section 5.2):

$$\text{acetyl ACP} + 7 \text{ malonyl-ACP} \longrightarrow \text{palmitoyl ACP} + 7\text{ACP} + 7\text{CO}_2$$
$$(6.9)$$

$$\text{palmitoyl ACP} + \text{malonyl ACP} \longrightarrow \text{stearoyl ACP} + \text{CO}_2 + \text{ACP}$$
$$(6.10)$$

6.7. NITROGEN ASSIMILATION

Plants obtain nitrogen by taking up nitrate and/or ammonium ions from the soil or through a symbiotic association with microorganisms that can utilize nitrogen gas from the air. There is a close association between photosynthesis and nitrogen metabolism, the photosynthetic light reactions providing the reductant both directly in the form of reduced ferredoxin and indirectly as NADPH and NADH for the conversion of nitrate to ammonia (Figure 6.6). In contrast to nitrite and ammonia, nitrate often accumulates in plant tissues suggesting that the activity of the enzyme nitrate reductase is rate limiting. In the field, the supply of nitrate is a major factor in determining the level of nitrate reductase activity and the rate of nitrate assimilation in green leaves (Hageman, 1979). There is also a close correlation between nitrate reductase activity and the level of reduced nitrogen present in the leaves, indicating that the enzyme is more likely to limit nitrate assimilation by the plant than lack of reductant generated by photosynthesis. Nitrate reductase is localized in the cytoplasm of the leaf cells and may be associated with the outer membrane of the chloroplast envelope. This enzyme produces nitrite from nitrate utilizing reductant in the form of NADH or NADPH

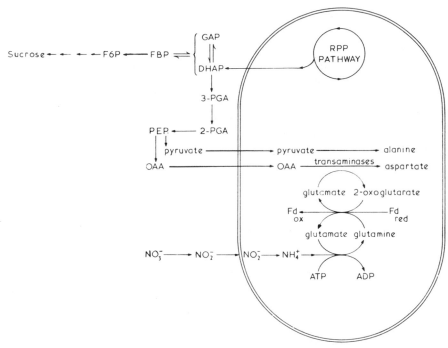

Figure 6.6.   The pathway of nitrate incorporation in leaves showing the reactions that occur in the chloroplast and cytosol.

[reaction (6.11)] although NADH appears to be the preferred reductant:

$$NO_3^- + NAD(P)H \longrightarrow Nad(P) + NO_2^- + H_2O \qquad (6.11)$$

Nitrate reduction in photosynthetic tissues is greatly enhanced by illumination, suggesting that the reductant derives indirectly from the chloroplast via one of the shuttle mechanisms that generate NADH in the cytoplasm (Chapter 5). Reducing power may also be obtained by mobilization of stored carbohydrate or by glycine oxidation in leaf mitochondria during photorespiration. Canvin and Woo (1979) have suggested that nitrate reductase and the mitochondrial electron transport chain compete for cytosolic NADH and that nitrate reduction can only occur in the dark when mitochondrial electron transport is inhibited (see also Section 5.9). Nitrate reductase is a complex enzyme incorporating molybdenum, FAD, and a cytochrome (b_{557}) in its structure. Nitrite, NO_2^-, generated by the activity of nitrate reductase, crosses the chloroplast envelope and is converted to NH_4^+ by the activity of nitrite reductase, which in leaves is an exclusively stromal enzyme. Nitrite reductase contains a sirohaem (iron–porphyrin) prosthetic group and a nonheme iron prophyrin center. The enzyme transfers six electrons to each NO_2^- ion as in reaction (6.12), utilizing reduced ferredoxin as the reducing agent:

$$NO_2^- + 6e^- + 8H^+ \longrightarrow NH_4^+ + 2H_2O \qquad (6.12)$$

Ammonia is converted to an organic form through the combined action of glutamine synthase and glutamate synthase (Miflin and Lea, 1976; 1977), as shown in Figure 6.6. There is probably very little free ammonia present in leaves as it is highly toxic to plant metabolism. Ammonia is incorporated into glutamine by the enzyme glutamine synthetase [reaction (6.13), which has a high affinity for ammonia $(K_m < 10^{-4} M)$]:

$$\text{L-glutamate} + NH_3 + ATP \underset{}{\overset{Mg^{2+}}{\rightleftharpoons}} 1 \text{ glutamine} + ADP + Pi$$
$$(6.13)$$

The enzyme, glutamate synthase (EC 1.3.7.1), is also called glutamine (amide):2-oxoglutarate amino-transferase (NADP oxidoreductase) (GOGAT) (Tempest et al., 1970). Glutamate synthase utilizes reduced ferredoxin (Lea and Miflin), 1974) in the following:

$$\text{L-glutamine} + 2\text{-oxoglutarate} \xrightarrow{\quad\substack{\text{reduced}\\ \text{ferredoxin}}\quad\substack{\text{oxidized}\\ \text{ferredoxin}}\quad} 2 \text{ L-glutamate} \qquad (6.14)$$

In contrast to glutamate synthases from bacteria roots and cotyledons, the chloroplast enzymes do not appear to utilize NAD(P)H. The leaf enzymes from spinach, *Vicia faba*, and maize show high affinities for their substrates and high maximal velocities (Wallsgrove et al., 1977; Tamura et al., 1980).

In the glutamine synthase–GOGAT system, glutamine acts as the acceptor for ammonia in the glutamine synthetase reaction as well as the product of assimilation in the glutamate synthase reaction (Figure 6.6).

6.8. OTHER COMPOUNDS

The biosynthetic pathway of chlorophyll and similar molecules from general metabolites such as glutamate occurs entirely within the plastids of greening plant cells. The chlorophylls and other biosynthetically related molecules such as hemes, siroheme, and porphyrins are tetrapyrroles. Many steps in the synthesis of chlorophyll are shared by other similar molecules, for example, vitamin B12, such that two or more products may be formed simultaneously. The porphyrin and phytol portions of the chlorophyll molecule appear to be synthesized separately. The synthesis of δ-amino-levulinic acid (ALA) has long been considered to be the first committed step in the biosynthesis of the porphyrin moiety but recent evidence suggests that one or more precursors of ALA may also be specifically committed to this pathway (Castelfranco and Beale, 1982). ALA appears to be synthesized from glutamate via 2-oxoglutarate and 2,5-dioxovalerate (Harel, 1978). Phorphobilinogen produced from ALA by the action of ALA dehydratase is converted to protoporphyrin IX into which Mg^{2+} is incorporated. The insertion of magnesium is the first unique step toward the chlorophyll branch of the pathway. The magnesium protoporphyrin IX is then methylated and in a subsequent series of reactions the methylated derivative is converted to protochlorophyllide. The photoreduction of protochlorophyllide to chlorophyllide can only be achieved if the protochlorophyllide is associated with a protein known as holochrome. These last steps involve reduction of the porphyrin to the chlorin state. Chlorophyll-*a* can then be synthesized by the addition of a polyisoprene-derived–long-chain alcohol to the tetrapyrrole. The protochlorophyllide–holochrome complex appears to be associated with the photosynthetic membranes as an extrinsic protein. During the final stages of biosynthesis the chlorophyll molecule becomes inserted into the membrane.

Carotenoid pigments are synthesized in the chloroplasts from mevalonic acid via geranylgeranyl pyrophosphate (Section 5.9). Light-dependent assimilation of sulfate into cysteine occurs in chloroplasts. The pathway requires ATP and reduced ferredoxin supplied by the light reactions. Sulfite reduction occurs via sulfite reductase, which is localized in the chloroplasts. The sulfide produced by this reaction is rapidly asimilated into cysteine by the action of cysteine synthase.

REFERENCES

Amir, J. and Preiss, J. (1982). *Plant Physiol.* **69**, 1027–1030.
Arnold, W. N. (1968). *J. Theoret. Biol.* **21**, 13–20.

Avigad, G. (1982). In *Encyclopedia of Plant Physiology*, New Series, Vol. 13A, Plant Carbohydrates I, Intracellular Carbohydrates (F. A. Loewus and W. Tanner, eds.), pp. 217–347. Springer-Verlag, Berlin.

Borovsky, D., Smith, E. E., and Whelan, W. J. (1976). *Eur. J. Biochem.* **62**, 307–312.

Canvin, D. T. and Woo, K. C. (1979). *Can. J. Bot.* **57**, 1155–1160.

Castelfranco, P. A. and Beale, S. I. (1982). In *The Biochemistry of Plants*, Vol. 8, Photosynthesis (M. D. Hatch and N. K. Boardman, eds.), pp. 376–423. Academic Press, New York.

Delrot, S. and Bonnemain, J. L. (1981). *Plant Physiol.* **67**, 560–564.

Downton, W. J. S. and Hawker, J. S. (1973). *Phytochemistry* **12**, 1551–1556.

Foyer, C. H., Rowell, J., and Walker, D. A. (1983). *Arch. Biochem. Biophys.* **220**, 232–238.

Giaquinta, R. (1977a). *Plant Physiol.* **57**, 750–755.

Giaquinta, R. (1977 b). *Nature* **267**, 369–370.

Giaquinta, R. (1979). *Plant Physiol.* **63**, 744–748.

Ghosh, H. P. and Preiss, J. (1965). *J. Biol. Chem.* **240**, 960–961.

Ghosh, H. P. and Preiss, J. (1966). *J. Biol. Chem.* **241**, 4491–4504.

Hageman, R. H. (1979). In *Nitrogen Assimilation in Plants* (E. J. Hewitt and C. V. Cutting, eds.), pp. 591–611. Academic Press, New York.

Harbron, S., Foyer, C., and Walker, D. A. (1981). *Arch Biochem. Biophys.* **212**, 237–246.

Harel, E. (1978). *Prog. Phyrochem.* **5**, 127–180.

Hawker, J. S. and Hatch, M. D. (1966). *Biochem. J.* **99**, 102–107.

Hawker, J. S., Ozbun, J. L., Ozakai, H., Greenburg, E., and Preiss, J. (1974). *Arch Biochem. Biophys.* **160**, 530–531.

Herold, A. (1980). *New Phytol.* **86**, 134–144.

Herold, A. and Lewis, D. H. (1977). *New Phytol.* **79**, 1–40.

Herold, A., Lewis, D. H., and Walker, D. A. (1976). *New Phytol.* **76**, 397–407.

Herold, A. and Walker, D. A. (1979). In *Membrane Transport in Biology* (G. Giebisch, D. C. Tosteson, and H. H. Ussing, eds.), pp. 411–439. Springer-Verlag, Berlin.

Huber, S. C. (1981). *Pflanzenphysiol.* **102**, 443–450.

Huber, S. C. and Israel, D. W. (1982). *Plant Physiol.* **69**, 691–696.

Huber, S. C. and Moreland, D. E. (1980). *Plant Physiol.* **65**, 560–562.

Huber, S. C. and Moreland, D. E. (1981). *Plant Physiol.* **67**, 163–169.

Huber, S. C. and Pharr, D. M. (1981). *Plant Physiol.* **68**, 1294–1298.

Kaiser, W. M. and Bassham, J. A. (1979). *Plant Physiol.* **63**, 109–113.

Kandler, O. and Hopf, H. (1982). In *Encyclopedia of Plant Physiology*, New Series, Vol. 13A, Plant carbohydrates I, Intracellular Carbohydrates (F. A. Loewis and W. Tanner, eds.), pp. 348–383. Springer-Verlag, Berlin.

Komer, E. (1982). In *Encyclopedia of Plant Physiology*, New Series, Vol. 13A, Plant carbohydrates I, Intracellular Carbohydrates (F. A. Loewus and W. Tanner, eds.), pp. 635–676. Springer-Verlag, Berlin.

Kuhn, D. H., Knauf, M. J., and Stumpf, P. K. (1981). *Arch. Biochem. Biophys.* **209**, 441–450.

Lea, P. J. and Miflin, B. J. (1974). *Nature* **251**, 614–616.

Leloir, L. F. and Cardini, C. E. (1955). *J. Biol. Chem.* **214**, 157–165.

Manners, D. J. and Sturgeon, R. J. (1982). In *Encyclopedia of Plant Physiology*, New Series, Vol. 13A, Plant Carbohydrates I, Intracellular Carbohydrates (F. A. Loewus and W. Tanner, eds.), pp. 635–676. Springer-Verlag, Berlin.

Miflin, B. J. and Lea, P. J. (1976). *Phytochemistry* **15**, 873–885.

Miflin, B. J. and Lea, P. J. (1977). *Ann. Rev. Plant Physiol.* **28**, 299–329.

Murphy, D. J. and Stumpf, P. K. (1981). *Arch. Biochem. Biophys.* **212**, 730–739.

Murphy, D. J. and Walker, D. A. (1982). *Planta* **156**, 84–88.

Pollock, C. J. (1976). *Plant Sci. Lett.* **7**, 27–31.

Pontis, H. G. (1977). In *International Review of Biochemistry*, Vol. 13 (D. H. Northcote, ed.), pp. 80–111. University Park Press, Baltimore.

Preiss, J. (1982). In *Encyclopedia of Plant Physiology*, New Series, Vol. 13A, Plant Carbohydrates I, Intracellular Carbohydrates (F. A. Loewus and W. Tanner, eds.), pp. 397–417). Springer-Verlag, Berlin.

Recondo, E. and Leloir, L. F. (1961). *Biochem. Biophys. Res. Commun.* **6**, 85–88.

Stitt, M. and ap Rees, T. (1979). *Phytochemistry* **18**, 1905–1911.

Stitt, M. and Heldt, H. W. (1981). *Plant Physiol.* **68**, 755–761.

Stitt, M., Mieskes, G., Soling, H. D., and Heldt, H. W. (1982). *FEBS Lett.* **145**, 217–222.

Stumpf, P. K. (1981). *Trend. Biochem. Sci.* **6**, 173–176.

Tamura, G., Kanki, M., Hirasawa, M., and Oto, M. (1980). *Agr. Biol. Chem.* **44**, 925–927.

Tempest, D. W., Meers, J. L., and Brown, C. M. (1970). *Biochem. J.* **117**, 405–407.

Thomas, D. R., Amiffin, A., Noh Hj Jalil, M., Yong, B. C. S., Cooke, R. J., and Wood, C. (1981). *Phytochemistry* **20**, 1241–1244.

Thomas, D. R., Noh Hj Jalil, M., Cooke, R. J., Yong, B. C. S., Ariffin, A., McNeil, P. H., and Wood, C. (1982). *Planta* **154**, 60–65.

Wallsgrove, R. M., Harel, E., Lea, P. J., and Miflin, B. J. (1977). *J. Exp. Bot* **28**, 588–596.

Weeden, N. F., Cseke, C., and Buchanan, B. B. (1982). *Plant Physiol.* **B59**, 336.

Zimmermann, G., Kelley, G., and Latzko, E. (1978). *J. Biol. Chem.* **253**, 5952–5956.

7

C4 METABOLISM

7.1. THE C4 CYCLE

The primary carboxylase in all photoautotrophs is ribulose-1,5-bisphosphate carboxylase. This, in collaboration with the other enzymes of the RPP pathway, provides the sole means of net carbon assimilation in photosynthesis. The first stable product of this reaction sequence is the C3 compound, PGA as shown originally in *Chlorella* by Calvin and colleagues. The experiments of Kortschak and later Hatch and Slack indicated that there were differences in the initial products of CO_2 fixation in sugarcane and maize from those described for *Chlorella*. Using $^{14}CO_2$ it was not possible to demonstrate early labeling of phosphoglyceric acid in sugarcane and maize, and instead malate was shown to be the primary stable product. It has since become clear that some plants, for example, maize and *Amaranthus* (Plate 7.1), possess an additional pathway of CO_2 uptake in which the first stable products of CO_2 assimilation are the C4 dicarboxylic acids, malate and aspartate (Hatch and Slack, 1970). These are therefore classified as C4 plants as opposed to C3 plants in which the primary product of CO_2 fixation is phosphoglycerate (Plate 7.1). Many C4 plants such as sugar cane, maize, and sorghum have high productivities and are of economic importance. C4 plants

Plate 7.1.   Examples of C4 plants (A) maize (*Zea mays*) (an NADP-malic enzyme type species).

show a low CO_2 compensation point and, unlike C3 plants, they do not loose a significant proportion of the carbon fixed in photosynthesis in the process of photorespiration, because the C4 cycle acts as a concentrating CO_2 pump.

Plants that exhibit C4 metabolism have two distinct types of cells which cooperate in net CO_2 fixation (Plate 7.2). These two cell types—the bundle sheath and mesophyll cells—are central to the operation of the C4 pathway and rapid metabolite transport occurs between the two. The mesophyll and bundle sheath cells are distinguished on the basis of different anatomy and their component enzyme activities for the assimilation of CO_2, NO_3^-, and

(b)

Plate 7.1. (*continued*) (B) *Amaranthus edulis* (an NAD-malic enzyme type species) and C3 plants.

SO_4^{2-} (Hatch, 1978; Black, 1973; Campbell and Black, 1982). The bio-chemistry of C4 photosynthesis has several variations but may be conveniently divided into two separate phases, a carboxylation stage that yields C4-dicarboxylic acids and a decarboxylation phase followed by fixation of the released CO_2 in the Calvin cycle. The C4 acid carboxylase and decarboxylase systems are spatially separated and are compartmentalized in the adjacent mesophyll and bundle sheath cells. CO_2 is transported from one to the other as the C4-dicarboxylic acids. In quantitative terms the intermediate pools of C4 acids in C4 photosynthesis are small (about 1 µequiv./g fresh weight) and their turnover rate is high ($t_{1/2} < 10$ sec). The initial carboxylation into C4 acids occurs in the mesophyll cells. These do not contain RuBP carboxylase and do not possess Calvin cycle activity, they do however contain the enzymes phosphoglycerate kinase and NADP–GAP dehydrogenase. The C4-dicarboxylic acids are transferred from the mesophyll cells to the bundle sheath cells (Figure 7.1) where they are decarboxylated. CO_2 can then be reassimilated via the reactions of the RPP pathway, which is localized in the bundle sheath cells.

(c)

Plate 7.1.   (*continued*) C3 plants (C) soya bean (*Glycine max*).

The vast majority of C4 plants have the Kranz-type of leaf anatomy (Plate 7.3), that is, radially arranged layers of bundle sheath and mesophyll cells. The possession of green bundle sheath cells alone does not characterize a C4 species; however, all plants with green bundle sheath cells surrounded by a single layer of green mesophyll cells typically exhibit C4 photosynthesis. A few species, for example, *Panicum miliodes*, are considered to be intermediate between the C3-type and the C4-type of photosynthesis.

The coordinated function of the mesophyll and bundle sheath cells is required to maintain the C4 cycle. Transport between the two types of cells may occur via the numerous plasmodesmata that form channels through the cell wall and are seen to connect the two cell types in electronmicrographs. The plasma membrane of each plant cell is continuous with the plasma

Plate 7.1.   (*continued*) (D) sunflower (*Helianthus annuus*).

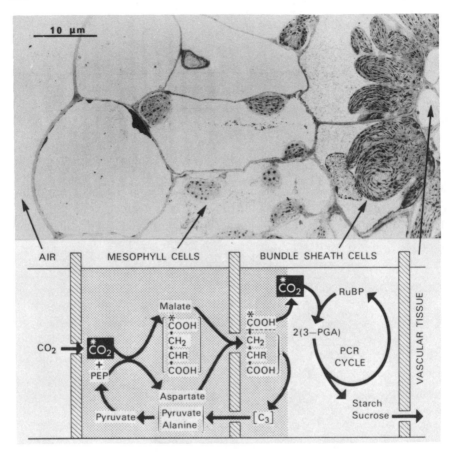

Plate 7.2. Light micrograph of *Panicum miliaceum* (an NAD-malic enzyme type species) leaf section showing the compartmentation of the C4 cycle, the carboxylation phase in the mesophyll cells, and the decarboxylation phase and the photosynthesis carbon reduction (PCF) cycle, which is the Calvin cycle, in the bundle sheath cells.

membrane of the adjacent cells and should, therefore, allow free diffusion of small molecules such as metabolites between the cells. Direct transport across the plasma membranes into the cell space and across the cell wall has been thought to be too slow to account for the rapid flux of C4 metabolites required for the operation of the cycle. The C4 cycle serves to transfer CO_2 from the atmosphere to the site of RuBP carboxylase in the bundle sheath cells and does not in itself contribute to net CO_2 fixation. It does, however, ensure an enrichment of the CO_2 concentration in the environment of RuBP

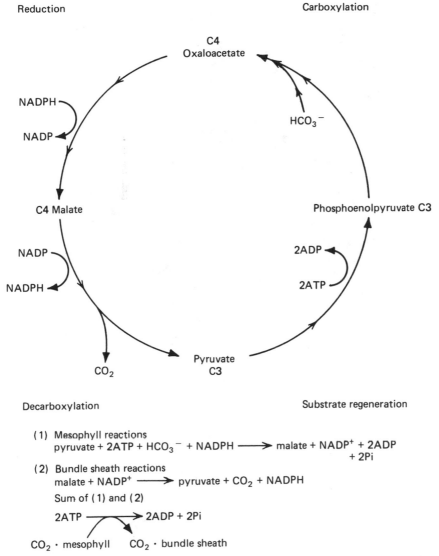

Reduction   Carboxylation

C4
Oxaloacetate

NADPH

NADP

C4 Malate

Phosphoenolpyruvate C3

NADP

2ADP

NADPH

2ATP

CO_2

Pyruvate
C3

Decarboxylation   Substrate regeneration

(1) Mesophyll reactions
pyruvate + 2ATP + HCO_3^- + NADPH $\longrightarrow$ malate + $NADP^+$ + 2ADP
$+ 2Pi$

(2) Bundle sheath reactions
malate + $NADP^+$ $\longrightarrow$ pyruvate + CO_2 + NADPH

Sum of (1) and (2)

2ATP $\longrightarrow$ 2ADP + 2Pi

$CO_2 \cdot$ mesophyll $CO_2 \cdot$ bundle sheath

Figure 7.1. A simplified diagram of the C4 cycle showing its function as an ATP-dependent CO_2 pump.

157

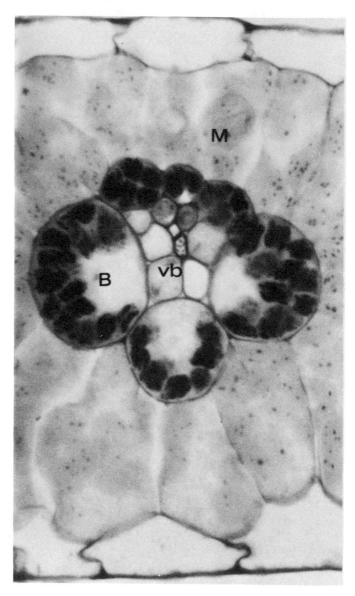

Plate 7.3. Light micrograph of a *Zea Mays* leaf section showing the central vascular bundle (vb) surrounded by the bundle sheath cells (B), whose chloroplasts contain heavily stained starch grains. Mesophyll cells (M).

carboxylase and may be regarded as an energy (ATP) dependent CO_2 pump (Figure 7.2).

Most C4 species are confined to areas of tropical and warm temperate climates, for example, the distribution of C4 monocotyledons in North America is positively correlated with higher summer temperatures although a few species extend into relatively cool northern regions (Teeri and Stowe, 1976). Similarly the distribution of C4 dicotyledons in North America is positively correlated with the potential evaporation in the summer months (Stowe and Teeri, 1978).

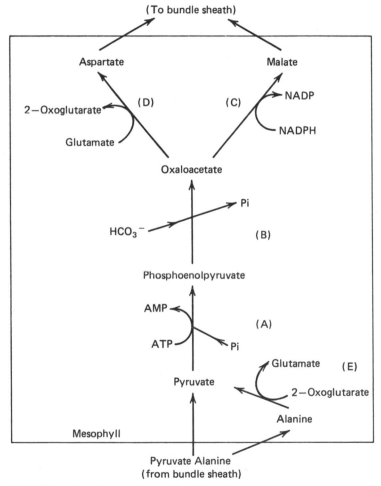

Figure 7.2. The carboxylation phase of the C4 cycle in the mesophyll cells. Reactions (A) and (C) occur in the chloroplasts, reaction (B), (D), and (E) in the cytosol.

C4 plants generally have a high temperature optimum (30–40°C) for photosynthesis. At temperatures below optimum the rate of photosynthesis is largely dependent on irradiance, which may not be totally saturating even at full sunlight. The rate of photosynthesis in most C4 species declines sharply at temperatures below about 10°C and many are killed by prolonged exposure to temperatures in the range of 0–10°C. This sensitivity to low temperatures may result from damage to both the soluble and membrane components of the chloroplast (Berry and Bjorkman, 1980). A small number of C4 plants do occur in cool-temperate climates, for example, *Spartina townsendii* and *Spartina anglica*. The C4 grass *S. townsendii* has been shown to have anatomy and gas exchange characteristics similar to those of C4 plants inhabiting hotter climates except that photosynthesis continues below 10°C at rates comparable with cool-temperate C3 plants. Similarly, photosynthetic carbon metabolism in *S. anglica*, which is abundant in cool-temperature habitats, has been shown to be similar to that of C4 plants of hotter climates (Smith et al., 1982).

7.2. PHOSPHOENOLPYRUVATE (PEP) CARBOXYLASE

In all C4 species the carboxylation reaction of the C4 cycle is facilitated by the action of the enzyme, PEP carboxylase (EC 4.1.1.31), which catalyzes

$$PEP + HCO_3^- + H_2O \longrightarrow oxaloacetate + Pi \qquad (7.1)$$

The enzyme is present in all higher plants and algae and also in some bacteria (O'Leary, 1982). Ting and Osmond (1973a) have suggested that at least four different forms of PEP carboxylase exist in higher plants. These are associated with different metabolic pathways and may be classified as a C3-photosynthetic PEP carboxylase, a C4-photosynthetic PEP carboxylase, a CAM–PEP carboxylase, and a dark or nonautotrophic enzyme. PEP carboxylase is of particular importance in the metabolism of C4 and CAM species. The reaction catalyzed by PEP carboxylase (7.1) is strongly exergonic $\Delta G^{\circ\prime} = -6$ to -8 kcal/mole and is essentially irreversible. It is clear that PEP carboxylase utilizes HCO_3^- as substrate, not CO_2. When assayed at pH 7.0 in the absence of inhibitors the enzymes from C4 species such as maize (Uedan and Sugiyama, 1976) and CAM species such as *Kalanchoe diagremontiana* have a higher affinity for HCO_3^- ($K_m < 20$ μM) than the enzymes from C3 species ($K_m = 100–300$ μM) (Ting and Osmond, 1973b).

The concentration of HCO_3^- in aqueous solution in equilibrium with air approaches 0.4 mM and is therefore in excess of the affinity of all enzyme forms under these conditions. The enzymes from all species require divalent metal ions for activity; they do not appear to bind a PEP–magnesium complex but bind each separately. At pH 8.0 in the presence of Mg^{2+} maize PEP carboxylase displays hyperbolic kinetics with respect to PEP, with a K_m of approximately 1 mM, however, the affinity of the enzyme for PEP is decreased with decreasing pH (Mukerji, 1977; Uedan and Sugiyama, 1976). At pH 7.0 the kinetics becomes sigmoidal. Glucose-6-phosphate activates the enzyme apparently by lowering the K_m for PEP, inducing hyperbolic reaction kinetics and also by decreasing the inhibitory action of malate. Malate and aspartate are powerful inhibitors of PEP carboxylase particularly at low pH, $[Mg^{2+}]$, or [PEP] but are less effective on enzymes from C3 species (Huber and Edwards, 1975a,b,c). Oxaloacetate is inhibitory to PEP-carboxylase activity but is only effective at nonphysiological concentrations.

The levels of PEP carboxylase are much higher in C4 plants than C3 plants such that the ratio of PEP carboxylase to RuBP carboxylase may be as high as 5:1 in C4 plants but is only approximately 0.1:1 in C3 plants. In all plants the enzyme appears to be localized exclusively in the cytosol of the cell (see also Section 8.4).

7.3.   THE CARBOXYLATION PHASE

The essential features of the carboxylation phase of the C4 cycle is shown in Figure 7.2. The reactions in the carboxylation phase that are compartmentalized in the mesophyll cells are similar in all C4 species and lead to the formation of malate and/or aspartate. In the basic C4 cycle the carboxylation phase relies on the production of PEP from pyruvate, which recycles either directly or indirectly from the bundle sheath cells. Pyruvate is imported into the mesophyll chloroplasts where PEP is formed by the action of the enzyme pyruvate, Pi dikinase (EC 2.7.9.1),

$$\text{pyruvate + ATP} \rightleftharpoons \text{PEP + AMP + PPi} \qquad (7.2)$$

This enzyme undergoes rapid light-mediated activation and dark-mediated inactivation in leaves and isolated mesophyll chloroplasts (Hatch, 1978). AMP inhibits the activation process, but in C4 plants the enzyme adenylate kinase (EC 2.7.4.3) has a primary role in effecting the rapid net conversion of AMP to ADP. C4 plants have been shown to contain high levels of adenylate kinase when compared with C3 plants and this activity is largely localized in the mesophyll chloroplasts. The general function of adenylate kinase is to maintain equilibrium concentrations of adenine nucleotides and the rate of conversion of ATP and AMP to ADP is similar to, or lower than, the rate of the reverse reaction. The maize mesophyll chloroplast enzyme converts ATP + AMP to ADP at a rate approximately fourfold that of the reverse reaction, thus mediating the net transfer of AMP to ADP following AMP production via pyruvate, Pi dikinase (Hatch 1982). The maize mesophyll chloroplasts contain large amounts of pyrophosphorylase to rapidly convert the pyrophosphate produced by the pyruvate, Pi dikinase, reaction to Pi (Figure 7.3).

In the chloroplast stroma the action of pyruvate, Pi dikinase, produces PEP [reaction (7.2)]. This must pass across the chloroplast envelope to the cytoplasm, which is the site of the primary carboxylase of the C4 cycle, that is PEP carboxylase. The action of PEP carboxylase [reaction (7.1)] produces oxaloacetate, which is transferred back to the chloroplast where it is converted into malate by the action of NADP–malate dehydrogenase (EC 1.1.1.82):

$$\text{oxaloacetate} + NADPH + H^+ \longrightarrow \text{malate} + NADP \qquad (7.3)$$

NADP–malate dehydrogenase is a light-modulated chloroplast enzyme that can be activated *in vitro* by thioredoxin via reduced ferredoxin and ferredoxin–

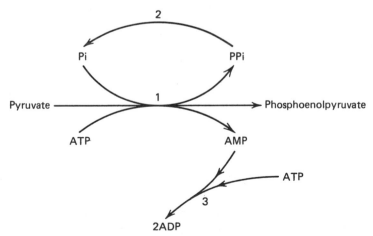

Figure 7.3. The action of (1) pyruvate Pi dikinase coupled to the activities of (2) pyrophosphatase and (3) adenylate kinase in the mesophyll chloroplast.

thioredoxin reductase (Schürmann and Jacquot, 1979) and also by a soluble protein factor that appears to act independently of the ferredoxin–thioredoxin system. Malate formed by reaction (7.3) can then be transported directly to the bundle sheath cells.

Some C4 species favor the production of aspartate rather than malate as the main export metabolite of the carboxylation phase in the mesophyll cells. Aspartate is produced in the cytoplasm as a result of the action of the enzyme aspartate aminotransferase (EC 2.6.1.1). This enzyme utilizes oxaloacetate produced by the action of PEP carboxylase in a transamination reaction with glutamate:

$$\text{oxaloacetate } + \text{ glutamate } \rightleftharpoons \text{ aspartate } + \text{ 2-oxoglutarate} \tag{7.4}$$

This enzyme is largely localized in the cytoplasm of the mesophyll cells of C4 species in which the decarboxylation phase is catalyzed by either NAD-malic enzyme or PEP carboxykinase. However, in species that utilize NADP-malic enzyme decarboxylation the aspartate aminotransferase is largely chloroplastic and much less aspartate is formed than in the other decarboxylation types.

In NAD-malic enzyme and PEP carboxykinase species the substrate pyruvate required for the pyruvate Pi-dikinase reaction may be produced largely from alanine, which is imported into the mesophyll cells from the bundle sheath cells. In the cytoplasm of the mesophyll cells alanine aminotransferase (EC 2.6.1.2) may be utilized to produce pyruvate:

$$\text{alanine } + \text{ 2-oxoglutarate } \rightleftharpoons \text{ pyruvate } + \text{ glutamate} \tag{7.5}$$

7.4. REGULATION OF THE CARBOXYLATION PHASE

In the conversion of pyruvate to malate [reactions (7.2)–(7.3)] two of the component enzymes pyruvate, Pi dikinase and NADP–malate dehydrogenase, are localized in the mesophyll chloroplast. The activities of both of these enzymes are modulated by light through a redox process possibly utilizing thioredoxin (Jacquot et al., 1981). However, the reductive modulation of pyruvate, Pi dikinase, is not considered to be of primary importance in the regulation of this enzyme. The regulation of pyruvate, Pi dikinase, is complex and is mediated by adenine nucleotides, Pi, and a high molecular weight protein factor (Sugiyama and Hatch, 1981). The protein factor is apparently required for interconversion of the active and inactive forms. Activation also requires Pi, Mg^{2+}, and reducing conditions and is inhibited by the presence of AMP. Inactivation of the enzyme apparently requires ADP (Edwards and Huber, 1982).

The activity of NADP–malate dehydrogenase in isolated intact mesophyll chloroplasts from *Zea mays* has been found to increase approximately 10-fold on illumination (Leegood and Walker, 1983). Light activation of the enzyme was found to be much diminished in the presence of the electron acceptors, PGA or oxaloacetate, which oxidize NADPH, rather than in the presence of pyruvate. Such results may be explained by effects on the redox state of the enzyme rather than direct effects of these substrates on the enzyme. Pronounced light activation of NADP–malate dehydrogenase is apparent in both leaves and chloroplasts but it is not known whether the reduced activity of this enzyme or pyruvate, Pi dikinase, at low light may become rate limiting for photosynthesis. The activity of both of these enzymes is reduced in some C4 species grown under weak illumination. However, at low light intensities the major limitation on photosynthesis is the rate of production of ATP and NADPH. At optimum temperatures the photosynthetic capacity of C4 plants is dependent largely on light intensity, the rate of photosynthesis increasing up to full sunlight. The quantum yield (moles of CO_2 fixed per mole of quanta absorbed) is similar for both C3 (19.1) and C4 (18.7) plants (Ehleringer and Bjorkman, 1977) under atmospheric conditions and temperatures between 20 and 25°C. Unlike C3 species C4 plants do not show a signficant O_2 inhibition of photosynthesis. Thus in C4 plants the quantum yield is similar at low (2%) O_2 and normal (21%) O_2 and is not effected by changes in temperature between 15 and 35°C as it is in C3 plants. C4 plants, however, require two more ATP molecules per CO_2 fixed, utilized to drive the C4 cycle, than C3 plants, which in theory require three ATP and two NADPH per CO_2. The high light saturation of C4 photosynthesis may therefore be linked to the requirement for assimilatory power. C4 photosynthesis requires a high ATP:NADPH ratio and the additional ATP may be generated by stimulation of cyclic and pseudocyclic electron flow (see Chapter 3).

In the mesophyll cells Pi is transported from the chloroplasts to the cytosol in the form of PEP and Pi must be returned to the chloroplast for continued CO_2 fixation. Pyruvate, Pi dikinase, is localized in the chloroplast and PEP carboxylase in the cytoplasm; carboxylation of a C3 precursor, therefore, requires the cooperative action of chloroplasts and cytoplasm. When isolated mesophyll chloroplasts are incubated with mesophyll cytoplasm, the addition of pyruvate and Pi in the light results in carbon assimilation (Edwards and Huber, 1978). Pyruvate and Pi must therefore be taken up by the chloroplasts and PEP must be exported for carboxylation to proceed. Transport of pyruvate is facilitated by a pyruvate translocator present on the mesophyll chloroplast envelope. In addition, a phosphate translocator exchanges PEP for Pi (Huber and Edwards, 1977) unlike the phosphate translocator of C3 plants. Together these translocators facilitate metabolite transport between the chloroplasts and cytoplasm in the carboxylation phase. Phosphoglycerate produced in the bundle sheath cells may be converted to triose phosphate in the mesophyll chloroplasts. The phosphate translocator on the mesophyll chloroplast envelope may be utilized to import cytosolic PGA in exchange for stromal triose phosphate. In C3 chloroplasts the export of triose phosphate in exchange for imported Pi links the activity of the Calvin cycle with the synthesis of sucrose. Sucrose synthesis occurs in the cytoplasm of the C4 mesophyll cells but is not directly linked to Pi–triose phosphate exchange across the chloroplast envelope as it is in C3 chloroplasts.

The allosteric regulation of PEP carboxylase activity by the products of the carboxylation phase, that is, oxaloacetate, malate, and aspartate, could prevent the accumulation of C4 acids and reduce the activity of the C4 cycle, in situations where the rate of the C4 cycle was substantially greater than that of the Calvin cycle. Activation of the enzyme by glucose-6-phosphate may prevent the loss of PEP from the C4 cycle, for example, through gluconeogenesis.

7.5. THE DECARBOXYLATION PHASE

Three pathways of C4-dicarboxylic acid decarboxylation are found in the bundle sheath cells of C4 species. These may be classified according to the enzyme utilized for the decarboxylation reaction.

NADP-Malic Enzyme

This is a chloroplast enzyme that requires Mg^{2+} or Mn^{2+} for activity. The pH optimum for activity increases from pH 7.4 to 8.5 with increasing malate concentration. Malate transported from the mesophyll to the bundle sheath cells is decarboxylated to pyruvate in the reaction:

$$malate \ + \ NADP \longrightarrow pyruvate \ + \ CO_2 \ + \ NADPH \qquad (7.6)$$

Bicarbonate is inhibitory to activity and this may serve to regulate the cycle and prevent excess CO_2 release in the bundle sheath cells. The activity of the Calvin cycle is linked to the activity of the NADP-malic enzyme through the NADP–NADPH pool. NADP-malic enzyme (EC 1.1.1.40) requires NADP and produces NADPH while the reactions of the Calvin cycle require NADPH and produce NADP (Figure 7.4). In NADP-malic enzyme species the chloroplasts are dimorphic. The bundle sheath chloroplasts are agranal whereas the mesophyll chloroplasts have grana. Thylakoids from the bundle sheath chloroplasts show low rate of electron transport from water to NADP because they are partially deficient in PSII. This results in decreased oxygen evolution and enhanced cyclic electron flow around PSI.

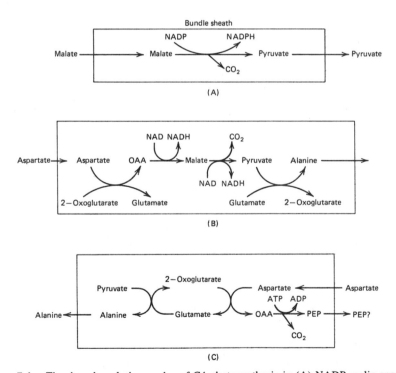

Figure 7.4. The decarboxylation cycles of C4 photosynthesis in (A) NADP-malic enzyme species, (B) NAD-malic enzyme species, and (C) PEP carboxykinase species.

In the Calvin cycle the refixation of each CO_2 molecule released from malate requires two NADPH and three ATP. One NADPH is provided by the NADP-malic enzyme reaction. The remaining NADPH may be provided by the limited noncyclic electron flow in the bundle sheath chloroplasts and also by shuttling PGA back to the mesophyll chloroplasts for conversion to triose phosphates. The C4 mesophyll chloroplasts contain enzymes of the reductive phase of the Calvin cycle allowing the conversion of PGA to triose phosphate, which could then return to the bundle sheath cells. Pyruvate produced by the decarboxylation of malate is transported back to the mesophyll cells to complete the cycle and maintain the carbon balance of the leaf. Examples of C4 species containing the NADP-malic enzyme system are *Zea mays* and *Sorghum bicolor* (Plate 7.1).

NAD-Malic Enzyme

This enzyme was first prepared from mitochondria by Macrae (1971) who demonstrated that this enzyme catalyzes the oxidative decarboxylation of malate to pyruvate:

$$NAD^+ + malate \rightleftharpoons pyruvate + CO_2 + NADH + H^+$$

$$(7.7)$$

The enzyme requires either Mg^{2+} or Mn^{2+} for activity. $NADP^+$ can replace NAD^+ although the latter is the preferred cofactor (Grover et al., 1981). NAD-malic enzyme (EC 1.1.1.39) is activated by acetyl CoA, CoA, FBP, fumarate, and sulfate (Grisson et al., 1983). This enzyme is the major if not the only decarboxylating enzyme in one group of C4 plants. In addition to its role in C4 metabolism, NAD-malic enzyme is also important in the intermediary metabolism of the plant cell since it allows the tricarboxylic acid cycle to operate in conditions in which pyruvate is not available from glycolysis. Similarly, the ability of this enzyme to perform carboxylation–decarboxylation reactions may be of importance in the regulation of intracellular pH. Enzymes that carry out reactions of this type have to be postulated to provide a buffering mechanism by regulating the levels of CO_2 in solution (Davies and Patil, 1973).

In NAD-malic enzyme species aspartate and malate produced from precursors in the mesophyll cells are transferred to the mitochondria of the bundle sheath cells where decarboxylation occurs. High levels of aspartate aminotransferase and alanine aminotransferase (EC 2.6.1.2) are present in both mesophyll and bundle sheath cells. Aspartate aminotransferase appears to be localized mainly in the mitochondria of the bundle sheath; the isoenzyme present in these cells is clearly distinct from that of the mesophyll cells. The action of aspartate aminotransferase in the bundle sheath produces oxaloacetate and glutamate [reaction (7.4)]. Oxaloacetate is subsequently reduced to malate via the action of NAD-malate dehydrogenase (EC 1.1.1.37):

$$\text{oxaloacetate} + \text{NADH} + \text{H}^+ \rightleftharpoons \text{malate} + \text{NAD} \qquad (7.8)$$

This mitochondrial enzyme provides malate for decarboxylation to pyruvate by the activity of NAD-malic enzyme [reaction (7.8)]. Pyruvate is then transferred to the cytoplasm where transamination with glutamate facilitated by the action of alanine aminotransferase produces alanine and 2-oxoglutarate [reaction (7.5)]. Alanine is subsequently transported to the mesophyll cells to complete the cycle. In NAD-malic enzyme and PEP-carboxykinase-containing species the major bundle sheath isoenzyme is distinct from the forms present in the mesophyll cells. The mitochondria of mesophyll and bundle sheath cells from NAD-malic enzyme species are dimorphic. The high flux of metabolites through the mitochondria in which decarboxylation through NAD-malic enzyme occurs has led to structural changes in the bundle sheath mitochondria, which are large and contain highly developed cristae. The bundle sheath mitochondria have high levels of NAD-malic enzyme. In *Panicum miliaceum* approximately 95% of the total NAD-malic enzyme is localized in the bundle sheath mitochondria. The mesophyll mitochondria from *P. miliaceum* have an NAD-malic enzyme activity similar to that reported for C3 mitochondria (Gardestrom and Edwards, 1983). Malate oxidation in the bundle sheath mitochondria was found to be insensitive to cyanide but inhibited by salicylhydroxamic acid. It has been shown that electron transport associated with malate oxidation occurs via the alternative nonphosphorylating oxidase pathway (Day et al., 1980). Oxidation through NAD-malic enzyme, in contrast to malate dehydrogenase, may be directly connected with an alternative pathway. Malate oxidation linked to an uncoupled, alternative electron pathway may allow decarboxylation to proceed without the restraints imposed by coupled electron flow through the cytochrome chain (Gardestrom and Edwards, 1983).

PEP Carboxykinase

This is a cytoplasmic enzyme (EC 2.6.1.1) that requires Mn^{2+} and ATP for activity. It is inhibited by dihydroxyacetone phosphate, PGA, and FBP. C4 plants, such as *Panicum maximum*, which contain this decarboxylating enzyme have low NADP-malic enzyme activity but have high levels of aspartate and alanine aminotransferases. In the mesophyll cells of these species the oxaloacetate produced by the carboxylation of PEP [reaction (7.1)] is largely converted into aspartate via transamination [reaction (7.4)]. Aspartate is imported into the bundle sheath cells where it is converted back to oxaloacetate via aspartate aminotransferase. PEP carboxykinase utilizes the oxaloacetate to produce CO_2 and PEP:

$$oxaloacetate + ATP \longrightarrow PEP + CO_2 + ADP \quad (7.9)$$

The fate of the PEP produced by this reaction is uncertain. PEP carboxykinase species will require two NADPH and between four and six ATP per CO_2 fixed in photosynthesis depending on the path of PEP. It is possible that the PEP is transferred directly back to the mesophyll cells, which would result in economy in ATP expenditure. However, the nitrogen balance must be maintained and it is possible that PEP is converted back to pyruvate in the bundle sheath and then to alanine. Alanine or glutamate can then be returned to the mesophyll cells to complete the cycle.

7.6. CARBOXYLATION IN THE CALVIN CYCLE

RuBP carboxylase (EC 4.1.1.39) catalyzes two competing reactions. RuBP can either be carboxylated to yield two molecules of PGA or oxygenated producing 2-phosphoglycollate and PGA. Although both the bundle sheath and mesophyll cells may possess the genes for RuBP carboxylase, differential gene expression occurs such that this enzyme activity in the C4 plant is compartmentalized solely in the bundle sheath chloroplasts (Link et al., 1978). Net carbon assimilation in C4 plants is achieved by spatial separation of C4 acid synthesis in the mesophyll cells and the RPP pathway in the bundle sheath cells. RuBP carboxylase in the C4 bundle sheath cells is similar to the enzymes from C3 species but it is considered that *in vivo* the oxygenase activity is kept significantly lower than that occurring in a C3 plant. It is suggested that the enriched CO_2 concentration generated by the C4 cycle in the bundle sheath favors the carboxylase reaction and depresses the oxygenase reaction. The C4 plants have not only evolved a CO_2 concentrating mechanism but have, in some cases, also achieved a decrease in the O_2 evolving capacity of the electron transport chain in the bundle sheath cells, as the chloroplasts of NADP-malic enzyme species show low PSII activity. The RuBP carboxylases from C4 species generally have a lower affinity for CO_2 than C3 species. C3 plants lack the ability to concentrate CO_2 at the site of carboxylation and this may have necessitated the evolution of carboxylases with high CO_2 affinity. The C4 carboxylases are able to remain fully active at high CO_2 concentrations that would inhibit C3 carboxylases.

The nature of the environment in the bundle sheath cells favors the suppression of photorespiration (Chapter 9). C4 plants have photorespiration but to a lesser extent than C3 plants. C4 plants have a lower compensation point (0–5 ppm CO_2) than C3 plants (35–70 ppm CO_2) thus C4 metabolism is saturated by air levels of CO_2. In addition the substrate for PEP carboxylase is HCO_3^- and there is no inhibition of this reaction by O_2. However, the additional assimilatory power required for total CO_2 fixation in the C4 plant can result in a lower photosynthetic efficiency than in the C3 plant under suboptimum environmental conditions. In warm climates where excess photosynthetic energy is available and CO_2 is limiting the C4 plant has the capacity to utilize more fully the potential of full sunlight, whereas C3 species can only photorespire faster.

As a result of the supression of photorespiration most C4 species produce much less glycollate, glycine, and serine than a typical C3 plant. The bundle sheath cells have the enzymes and potential for photorespiration. Glycollate metabolism in C4 plants is not linked to the RPP pathway as it is in C3 plants and appears to be shared between the mesophyll and bundle sheath cells. Glycollate is produced in the bundle sheath where the enzymes of the glycollate pathway metabolize glycollate to glycerate. Glycerate kinase is localized in mesophyll chloroplasts suggesting that glycerate formed in the bundle sheath is transported to the mesophyll chloroplasts where its metabolism may be directly linked to sucrose synthesis.

7.7. THE BENEFITS OF C4 PHOTOSYNTHESIS

The operation of the C4 cycle requires two ATP molecules per CO_2 fixed in addition to the three ATP and two NADPH required to drive the Calvin cycle. Photosynthesis in C4 plants is at the best advantage at high light intensities, where ATP production is not limited by light, and at high temperatures where C3 plants suffers from photorespiration and water loss through transpiration. At 2% O_2, C3 plants have a significantly lower quantum yield (36%) than C4 species when measured at 30°C under low light. With increasing $[O_2]$ the quantum efficiency is decreased in C3 plants because of O_2 inhibition of photosynthesis, which is not observed in C4 species. At 30°C, atmospheric CO_2 and 21% O_2, the measured quantum yields of C3 and C4 are comparable. In warmer climates conditions promote the depletion of CO_2 in the air of the environment surrounding the leaf and cause the stomata to partially close to reduce water loss (Bjorkman, 1976). These conditions hinder CO_2 penetration into the leaf and render C3 plants under severe photorespiratory stress because CO_2 would become limiting and O_2 inhibition of photosynthesis would be increased. C4 plants can maintain photosynthesis at a reduced stomatal conductance and thereby conserve water. The low K_m of PEP carboxylase for HCO_3^- and low compensation point provide a highly efficient trapping system for CO_2. Any CO_2 lost in the photorespiratory pathway may be efficiently retrieved by the action of PEP carboxylase. Under temperate conditions the advantages of C4 over C3 plants are negligible. C4 species are generally not cold tolerant and growth rates may fall below that of C3 species in areas where the temperature fluctuates over a growing season. C3 plants usually photosynthesize better and also grow better at low temperatures than do C4 plants. At lower temperatures CO_2 is not generally rate limiting nor is O_2 inhibition of photosynthesis as serious to C3 plants as it is at high temperatures. C4 photosynthesis may be limited at low temperatures because of cold liability of certain of the component enzymes; for example, there is a large increase in the energy of activation of PEP carboxylase below 10°C in some C4 species (Edwards and Huber, 1982).

C3 plants may have as much as 5 times the RuBP carboxylase content as C4 species. The high percentage of the total soluble protein occupied by this enzyme is necessary in C3 plants in order to trap CO_2 from the atmosphere. C3 species, therefore, expend more nitrogen in the synthesis of this enzyme than do C4 plants. C4 plants have been shown to attain maximum growth rates at lower soil nitrogen concentrations than C3 species and contain less total leaf nitrogen (Brown, 1978). These results would suggest that C4 plants utilize nitrogen more efficiently than do C3 species. Nitrogen assimilation appears to be largely compartmentalized in the mesophyll cells, which contain the majority of the glutamine synthetase, nitrate, and nitrite reductases.

REFERENCES

Berry, J. and Bjorkman, O. (1980). *Ann. Rev. Plant Physiol.* **31**, 491–543.

Black, C. C. (1973). *Ann. Rev. Plant Physiol.* **24**, 253–286.

Bjorkman, O. (1976). In *CO₂ Metabolism and Plant Productivity* (R. H. Burris, and C. C. Black, eds.), pp. 287–309. University Park Press, Baltimore.

Brown, R. H. (1978). *Crop. Sci.* **18**, 93–98.

Campbell, W. H. and Black, C. C. (1982). In *Cellular and Subcellular Localization in Plant Metabolism.* Recent advances in Phytochemistry, Vol. 16 (L. L. Creasy and G. Hrazdina, eds.), pp. 223–248. Plenum Press, New York.

Davies, D. D. and Patil, K. D. (1973). *Symp. Soc. Exp. Biol.* **22**, 513–543.

Day, D. A., Avron, G. P., and Laties, G. G. (1980). In *The Biochemistry of Plants*, Vol. 2, pp. 197–241. Academic Press, New York.

Edwards, G. E. and Huber, S. C. (1978). In *Proceedings of the Fourth International Congress on Photosynthesis* (D. O. Hall, J. Coombs, and T. W. Goodwin, eds.), pp. 95–106. The Biochemical Society, London.

Edwards, G. E. and Huber, S. C. (1982). In *The Biochemistry of Plants*, Vol. 8, Photosynthesis (M. D. Hatch and N. K. Boardman, eds.), pp. 238–282. Academic Press, New York.

Ehleringer, J. and Bjorkman, O. (1977). *Plant Physiol.* **59**, 86–90.

Grover, S. D., Canellas, P. F., and Wedding, R. T. (1981). *Arch. Biochem. Biophys.* **209**, 396–407.

Grisson, C. B., Canellas, P. F., and Wedding, R. T. (1983). *Arch. Biochem. Biophys.* **220**, 133–144.

Hatch, M. D. (1978). In *Current Topics in Cellular Regulation 14* (B. L. Horecher and E. R. Stradtman, eds.), pp. 1–27. Academic Press, New York.

Hatch, M. D. (1982). *Aust. J. Plant Physiol.* **9**, 287–296.

Hatch, M. D. and Slack, C. R. (1970). *Ann. Rev. Plant Physiol.* **21**, 141–162.

Huber, S. C. and Edwards, G. E. (1975a). *Plant Physiol.* **55**, 835–844.

Huber, S. C. and Edwards, G. E. (1975b). *Plant Physiol.* **56**, 324-331.

Huber, S. C. and Edwards, G. E. (1975c). *Can. J. Bot.* **53**, 1925–1933.

Huber, S. C. and Edwards, G. E. (1977). *Biochim. Biophys. Acta* **462**, 583–612.

Jacquot, J.-P., Buchanan, B. B., Martin, F., and Vidal, J. (1981). *Plant Physiol.* **68**, 300–304.

Leegood, R. C. and Walker, D. A. (1983). *Plant Physiol.* **71**, 513–518.

Link, G., Coen, D., and Bogorad, L. (1978). *Cell* **15**, 725–731.

Macrae, A. R. (1971). *Biochem. J.* **122**, 495–501.

Mukerji, S. K. (1977). *Arch. Biochem. Biophys.* **182**, 352–359.

O'Leary, M. (1982). *Ann. Rev. Plant Physiol.* **33**, 279–315.

Schurmann, P. and Jacquot, J.-P. (1979). *Biochim. Biophys. Acta* **569**, 309–312.

Smith, A. M., Woolhouse, H. W., and Jones, D. A. (1982). *Planta* **156**, 441–448.

Stowe, L. G. and Teeri, J. A. (1978). *Am. Nat.* **112**, 609–623.

Sugiyama, T. and Hatch, M. D. (1981). *Plant and Cell Physiol.* **22**, 115–126.

Teeri, J. A. and Stowe, L. G. (1976). *Oecologia* **23**, 1–12.
Ting, I. P. and Osmond, C. B. (1973a). *Plant Physiol.* **51**, 448–453.
Ting, I. P. and Osmond, C. B. (1973b). *Plant Physiol.* **51**, 439–447.
Uedan, K. and Sugiyama, T. (1976). *Plant Physiol.* **57**, 906–910.

8

CRASSULACEAN ACID METABOLISM (CAM)

8.1. CHARACTERISTICS OF CAM PLANTS

The term CAM is derived from the family name Crassulaceae in which a diurnal fluctuation in the malic acid content was first observed. Crassulacean acid metabolism has now been clearly shown to occur in several phyla including approximately 25 angiosperm families, one known gynosperm, and two ferns. The Crassulaceae is a large and geographically widespread family and contains genera such as *Sedum* and *Kalanchoe* (Plate 8.1). CAM is an adaptation to arid environments where water conservation is of prime

(a)

Plate 8.1.   (A) *Kalanchoe daigremontiana.*

importance (Osmond, 1978). CAM plants generally reside in habitats where they must withstand diurnal and seasonal periods of alternating wet and very dry conditions. Plants that exhibit this type of photosynthetic process are generally of tropical origin (mainly southern Africa) where conditions can be hot, xeric, and dry.

For many decades CAM has been equated with succulence. Perennial species that have greatly thickened vegetative organs modified to store large amounts of water are often described as succulent (Plate 8.2). The occurrence of succulence and the CAM mode of photosynthesis together is common but not mutually exclusive. CAM plants are protected by a well-developed cuticle to maximize resistance to water loss. Stomata are found on all shoot surfaces but at very low densities. The photosynthetic tissue (chlorenchyma) is composed of large vacuolated cells with numerous large chloroplasts. The CAM cell is characterized by the presence of a large vacuole that can occupy over 90% of the cell volume and is used to store acids, Pi, and

Plate 8.1.   (*continued*) (B) *Kalanchoe pumila*.

water. Each night CAM requires the temporary sequestration of large amounts of malic acid away from other metabolic processes. The vacuole of the chlorenchyma cell is one of the sites utilized for this purpose. The vacuole has been shown to be the major storage site for malate, isocitrate, and Pi. The vacuolar pools of malate and Pi change over the 24-hr CAM cycle while the isocitrate pool remains constant. When tissue from CAM species is fed $^{14}CO_2$ at night, the ^{14}C is incorporated into malate and sequestered in the vacuole (Plate 8.3). In the following light period the ^{14}C-malate in the vacuole is depleted and ^{14}C is incorporated into the products of the RPP pathway.

The stomata of CAM species open mainly at night during which time CO_2 assimilation occurs but water loss due to evaporation is minimal. During

(a)

Plate 8.2.   Examples of succulent plants (A) *Echeveria agavoides.*

daylight hours when evaporative demand would be maximal the stomata are closed. Storage carbohydrate is converted stoichiometrically to malic acid during CO_2 uptake in the dark, the fixation of CO_2 (as HCO_3^-) is facilitated by the action of PEP carboxylase and results in the accumulation of malic acid. In the light when net CO_2 exchange with the environment is negligible, malate is decarboxylated, CO_2 is refixed by the action of the RPP pathway, and storage carbohydrate is resynthesized. In this manner the synthesis and degradation of malate and starch shows a reciprocal diurnal fluctuation reflecting daily cycles of carbohydrate and acid metabolism. In some CAM species there is a damped but persistant circadian rhythm of changes of flux through PEP carboxylase in plants kept in continuous darkness and at constant temperatures (Warren and Wilkins, 1961). Similarly, in continuous light, rhythms of CO_2 uptake and water loss corresponding to the normal dark period have been observed (Luttge and Ball, 1978). These circadian rhythms appear to be limited by physiological processes such as

(b)

Plate 8.2.   (*continued*) (B) *Haworthia truncata* and *Lithops schwantesii* var. triebneri.

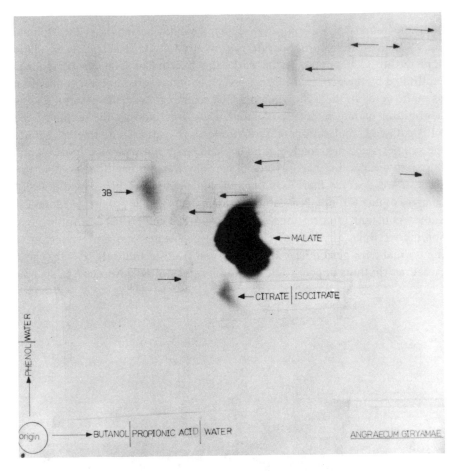

Plate 8.3. Radioautograph of ethanol extract of leaves of the orchid *Angraecum gyriamae* incubated in darkness overnight [Cockburn, W. and Thurston, P. A. (1977), unpublished].

stomatal conductance and tonoplast fluxes and their relevance in the normal diurnal cycle has yet to be understood.

Studies on the carbon assimilation pathways in CAM plants have been greatly facilitated by the discovery of inducible CAM species such as the annual succulent *Mesembryanthemum crystallinium*. In this plant a CAM mode of carbon assimilation can be induced by gradually increasing the level of plant water stress as the life cycle progresses (Winter and Von Willet, 1972). In *M. crystallinium* the induction of CAM is accompanied by a net synthesis of the enzyme, PEP carboxylase. Similarly, CAM may be induced in some species in response to changes in the photoperiod as a seasonal adaptation to drought. In this way short days favor CAM induction in *Kalanchoe blossfeldiana* such that CAM is absent during 16 hr/8 hr or 14 hr/10 hr day/night cycles and promoted by 12 hr/12 hr schedules and even more by 9 hr/15 hr regimes (Queiroz and Brulfert, 1982). There are, therefore, some succulent species that are of the C3-type when grown in favorable conditions but shift to CAM when stressed and in most cases this shift is reversible when suitable conditions return.

There are many similarities in the biochemistry of CAM and C4 photosynthesis. In both, C4-dicarboxylic acids are synthesized via PEP carboxylase as intermediates carrying CO_2 from the environment to RuBP carboxylase. In contrast to C4 metabolism CAM species maintain a large dicarboxylic acid pool (approximately 200 μequiv./g fresh weight) and turnover is relatively slow ($t_{1/2} > 10^4$ sec). CAM cells utilize 15–20% of their total organic material in the daily operation of the CAM cycle (Black et al., 1982). CO_2 is released from the C4 dicarboxylic acids by one of three alternative dicarboxylase systems based on NAD or NADP-malic enzyme or PEP carboxykinase as it is in C4 plants. In all cases this results in an enhanced CO_2 concentration in the vicinity of RuBP carboxylase. In the CO_2-concentrating mechanism of CAM, PEP-carboxylase and the decarboxylase systems are localized within the same cell but the phases of C4 acid synthesis and decarboxylation are

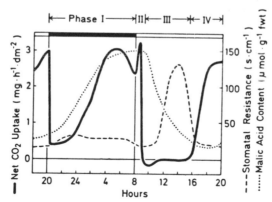

Figure 8.1.   Diurnal fluctuations in net CO_2 exchange (solid line), malic acid rhythm (dotted line), and stomatal resistance (broken line) during the phases of CAM in *Katanchoe daigremontiana*. The experimental conditions were 15°C night, 25°C day, and light intensity 60 W/m². [Taken from Kluge, M., Fischer, A., and Buchanan-Bollig, I. C. (1982). In *Crassulacean Acid Metabolism* (I. P. Ting and M. Gibbo, Eds.), p. 32. Waverly Press, Baltimore].

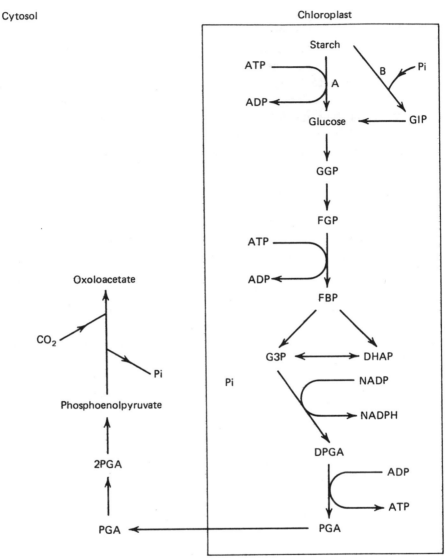

Figure 8.2. Diagram to show the path of carbon mobilization from starch via α-amylase and (A) glucokinase or (B) phosphorylase and glycolysis in CAM plants.

separated in time. This requires complex control of the alternative metabolic pathways in response to environmental signals such as light and water stress. Concentration of CO_2 in the cell during the day is achieved by a high rate of C4 decarboxylation releasing CO_2 and also resulting in stomatal closure. The overall pattern of carbon flow is one in which nocturnally synthesized malic acid serves as a temporary store for CO_2. This is depleted the following day by decarboxylation and refixation of the CO_2 occurs via the Calvin cycle. The malate store is again replenished during the night. Endogenous storage of CO_2 enables the plant to keep photosynthesis in operation behind closed stomata allowing little water loss and preventing loss of thylakoid function through photoinhibition and carbon loss through photorespiration. The leaves of C3 and C4 plants are subject to photoinhibitory damage when they are illuminated at full sunlight in the absence of external CO_2 in conditions that prevent the internal generation of CO_2 in significant amounts (Powles et al., 1980). Osmond et al. 1980 have suggested the recycling of respiratory CO_2 via the CAM cycle and the internal release of CO_2 via photorespiration, serve as significant internal sources of CO_2 in CAM plants that effectively protect against photoinhibition. In support of this suggestion Osmond (1982) has shown that after as long as 6 months in air, without water and with closed stomata, illuminated cladodes of CAM plants show no change in the low temperature fluorescence emission associated with PSII, reflecting the absence of photoinhibition.

The diurnal CAM cycle may be seen as composing of four typical phases (Figure 8.1) distinguishable by their gas exchange characteristics (Wilkins, 1969; Kluge and Ting, 1978). The gas exchange pattern of a given CAM plant may be significantly modulated by external factors but the pattern shown in Figure 8.1 may be taken as a model for the succession of CAM phases 1–4 as suggested by Osmond (1978).

8.2. PHASE 1—NOCTURNAL CO₂ FIXATION AND MALIC ACID SYNTHESIS

Phase 1 occurs at night and is a period of CO_2 assimilation. CO_2 from the atmosphere together with CO_2 supplied endogenously by respiration is fixed by the action of PEP carboxylase [reaction (7.1)] and the oxaloacetate thus formed is converted to malate by the action of NAD–malate dehydrogenase [reaction (7.8)]. During the phase of nocturnal CO_2 fixation PEP carboxylase is highly active, the activities of this enzyme and NAD–malate dehydrogenase are much greater than required to support the observed rates of malic acid synthesis. The PEP utilized in CO_2 fixation is generated from storage carbohydrates via glycolysis (Figure 8.2). Oxaloacetate resulting from the β-carboxylation of PEP is reduced to malic acid in a reaction that can be coupled to NADH production by the glyceraldehyde phosphate dehydrogenase reaction in such a way that malate synthesis from glucan is energetically self-sustaining. The rate and extent of malic acid synthesis is governed both by external factors such as temperature and the light regime and by internal factors such as the water potential of the tissue and the ability of the tonoplast to act as a sink for malic acid. Field studies on cacti, for example, have shown that net CO_2 assimilation or acid accumulation at night is approximately 90% saturated when the total day time photosynthetically active radiation incident on stems is between 20 and 40 mole/m²/day (Nobel and Hartsock, 1983). Malate produced by nocturnal CO_2 fixation is transported into the vacuole for overnight storage. It is likely that the tonoplast membrane contains

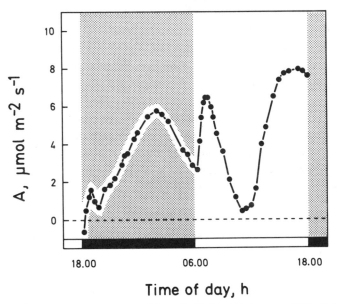

Figure 8.3. Change in the rate of net CO_2 assimilation (A) during a 12-hr dark/12-hr light cycle in a fully expanded attached leaflet of *Kalanchoe pinnata*. PAR (400–700 nm) was 800 μeinstein/m² · sec. Leaf temperature, leaf-air vapor pressure difference, and ambient partial pressure of CO_2 were constant at 20°C, 9–10 mbar and 32 μbar, respectively [see also Winter, K. (1980). *Plant Physiol.* **66**, 917–921].

an ATPase for pumping protons into the vacuole. This may be coupled either directly or indirectly to a malate^{2-} exchange carrier as has been shown to be present on tonoplast membranes from *Kalanchoe daigremontiana* (Buser-Suter et al., 1982a,b). Malate^{2-} could bind to this carrier on the cytoplasmic face and subsequently be removed from the exchange equilibrium of the carrier on the vacuolar face because of the dissociation equilibria at vacuolar pH values that would favor H-malate and H$_2$-malate. The undissociated acid may diffuse back to the cytoplasm, especially toward the end of the dark period as the vacuole becomes increasingly more acid. In the cytoplasm it would again dissociate to malate^{2-} and 2H$^+$. A K$^+$/H$^+$ exchange mechanism may also exist at the tonoplast such that K$^+$ is extruded from the vacuole as a consequence of H$^+$ pumping. Accumulation of malic acid in the vacuole may be increasingly limited in the dark by the availability of ATP required by the tonoplast ATPase.

The carbohydrate source of nocturnal malic acid synthesis is often found to be starch. In *Bryophyllum tubiflorum* and *K. daigremontiana* starch has been shown to account for about two thirds of the carbon while in *Opuntia aurantiaca* less than 40% of the carbon is derived from starch. In these species the pool of glucan (low molecular weight polymers of glucose) may be sufficient to account for the remaining carbon required. A few CAM species such as pineapple may use glucose and fructose as the major carbohydrate source supplemented with starch and/or glucans. In glycolysis the ATP requiring enzyme phosphofructokinase catalyses the first unique reaction of the pathway and is considered to be the major site of regulation of glycolytic carbon flow (Turner and Turner, 1980) for several reasons. Under physiological conditions ATP phosphofructokinase catalyzes an essentially irreversible reaction, which is found to be far from equilibrium *in situ*. Phosphofructokinase is modulated by a number of positive and negative effects such as PEP and shows complex kinetics. The activity of the enzyme is found to change in a similar fashion to induced changes in the overall rate of glycolysis (Osmond and Holtum, 1982). CAM plants tend to have higher activities of phosphofructokinase than do C3 and C4 species.

A PPi-dependent phosphofructokinase that catalyses reaction (8.1) was first shown to be present in pineapple leaves (Carnal and Black, 1983):

$$F6P \ + \ PPi \ \rightleftharpoons \ FBP \ + \ Pi \qquad (8.1)$$

This enzyme is cytosolic in origin and is considered to be important in the conversion of soluble sugars to trioses for PEP synthesis. It is widely distributed among photosynthetic organisms (Black et al., 1982) and may serve as an alternative to ATP-dependent phosphofructokinase in plant metabolism. PPi phosphofructokinase appears to present in almost all CAM plants at activities from 4- to 70-fold greater than the ATP phosphofructokinase (Carnal and Black, 1983). Fructose-2, 6-bisphosphate is an activator of PPi phosphofructokinase in C3, C4, and CAM species and the enzyme is not detectable in spinach leaves (Cseke et al., 1982) or in some CAM species (Carnal and Black, 1983) unless the activator fructose-2,6-bisphosphate is present. The presence of the PPi-dependent phosphofructokinase in the cytoplasm and the ATP-dependent phosphofructokinase in both the chloroplasts and cytoplasm (Kelly and Latzko, 1976) suggests the possibility that two glycolytic pathways may operate in green cells.

8.3. PHASE 2—THE BURST OF CO_2 UPTAKE UPON ILLUMINATION

Following the period of nocturnal CO_2 fixation the onset of illumination stimulates a burst in CO_2 uptake that can last for up to 2 hr (Figure 8.3). The duration of this morning burst is equivalent to phase 2 of the diurnal CAM cycle. The light-stimulated burst in CO_2 uptake can be observed both under natural conditions where irradiance gradually increases at dawn and also under laboratory conditions where plants are subjected to an abrupt transition from dark to light. Phase 2 links the nocturnal phase of malic acid synthesis and storage to the phase of malic acid mobilization and decarboxylation in the light. In phase 2 the processes of CO_2 fixation mediated by PEP-carboxylase and malic acid synthesis occur simultaneously with decarboxylation of malic acid and CO_2 fixation in the RPP pathway. When malic acid is released from the vacuole to the cytoplasm it is decarboxylated by either NAD-malic enzyme [reaction (7.7)], NADP-malic enzyme [reaction (7.6)], NAD–malate dehydrogenase [reaction (7.8)], or by PEP carboxykinase [reaction (7.9)]. The intracellular partial pressure of CO_2 may be as high as 10,000 μbar during deacidification favoring the carboxylation of RuBP and suppressing oxygenation. However, the decarboxylation reactions could also provide substrates for PEP carboxylase; for example, the PEP-carboxykinase reaction (7.9) leads to stoichiometric production of PEP in the cytoplasm. NADP-malic enzyme decarboxylation also leads to the formation of considerable amounts of PEP, since PEP is produced by the action of pyruvate-Pi dikinase from pyruvate [reaction (7.2)] in the chloroplasts and is exported to the cytoplasm to be further metabolized. PEP-carboxylase activity is, however, suppressed in the light and this is gradually achieved as phase 2 progresses. Throughout phase 2 the overlapping processes of carboxylation through PEP carboxylase and RuBP carboxylase change relative to one another in a continuous shift from a phase 1 path of carbon flow at the beginning to a phase 3 type path at the conclusion. The mainstream of carbon flow is therefore not constant during phase 2. The regulatory properties limiting the activities of the primary carboxylation reactions through PEP carboxylase and RuBP carboxylase are of primary importance in determining the course of this shift. Winter and Tenhunen (1982) have shown that, during the morning burst of CO_2 uptake in leaves of *K. daigremontiana*, CO_2 fixation occurs simultaneously via PEP carboxylase and RuBP carboxylase with the former being the predominant path of CO_2 fixation in the early part of phase 2 and the latter progressively achieving dominance as phase 2 comes to an end. When the morning burst of CO_2 fixation reached a maximum value, the onset of malic acid decarboxylation was initiated and may have been instigated by malic acid release from the vacuoles (Winter and Tenhunen, 1982). The subsequent decline in net CO_2 uptake from the air probably results from stomatal closure induced by the increase in the intercellular CO_2 partial pressure (Cockburn et al., 1979) and also by inhibition of PEP carboxylase by the increased cytoplasmic concentration of malic acid (Winter, 1982a). These results confirm earlier observations that ^{14}C-malate is a main product of [14]CO_2-fixation during the morning burst (Osmond and Allaway, 1974) and also the demonstration of CO_2 fixation via RuBP carboxylase particularly during the closing stages of phase 2 (Kluge et al., 1982). The relative distribution of carbon between the two pathways during phase 2 is also dependent on light intensity and temperature. High light intensities and optimum temperatures (~30°C) favor malic acid decarboxylation and can cause an earlier onset of this process. In phase 2, PEP carboxylase may operate at the expense of carbon skeletons derived from the Calvin cycle.

During phase 2 the vacuole changes its storage behavior becoming a source of malic acid instead of a sink. This change is not initiated by a light stimulus. At the beginning of phase 2 the vacuole still acts as a sink for malic acid. Osmond and Allaway (1974) showed a long half-lifetime for malate synthesized during initial phase 2 and malic acid accumulation may be maintained at this time. It is not known how the capacity of the vacuole to store malic acid is regulated. When the maximum capacity of malic acid storage is not reached during nocturnal CO_2 fixation, then phase 2 is extended and the onset of interconversion in the dominance between the two pathways is retarded.

8.4. THE REGULATION OF PEP CARBOXYLASE IN CAM

The regulation of PEP carboxylase is an important factor in determining the path of carbon flow in the CAM cycle. The regulatory properties of the enzyme undergo characteristic changes through the diurnal CAM cycle. PEP carboxylase may exist in kinetically distinct forms during the day and night. PEP carboxylase from plants such as *K. blossfeldiana* and *Mesembryanthum crystallinium* have a much lower K_m for PEP, a much higher K_i for malate and a lower K_a for glucose-6-phosphate during the night than during the day. Such changes in properties would favor fixation by PEP carboxylase at night and prevent it in the day, thus preventing futile competition with the C3 cycle. The PEP carboxylases from CAM plants have a higher affinity for PEP and a higher V_{max} than PEP carboxylases from C3 and C4 species. In all PEP carboxylases malic acid is a potent competitive feedback inhibitor, particularly at acid pH. This inhibition can be reversed by glucose-6-phosphate, which increases the affinity of the enzyme for PEP without changing V_{max}. The most widely accepted theory of the regulation of PEP carboxylase activity in CAM species is central on feedback inhibition by malate such that the activity of the enzyme [reaction (7.1)] is suppressed during the light by malate released from the vacuole. Recent studies indicate that the enzyme undergoes marked changes in its sensitivity to inhibition by malate and in its affinity for PEP during the CAM cycle. PEP carboxylase from *M. crystallinium* apparently exists in two different kinetic forms, in the light and in darkness, which differ in their capacity for malic acid synthesis. In the daytime the sensitization of the enzyme to malate inhibition and the decreased affinity for PEP limit CO_2 fixation by PEP carboxylation (Von Willert et al., 1979) and favor CO_2 fixation by RuBP carboxylase. Malic acid apparently acts as a potent regulator of PEP carboxylase in the light by direct inhibition and possibly also by maintaining the highly malate-sensitive kinetic state of the enzyme. This is particularly effective at lower pH values such as may occur when malic acid is released to the cytoplasm. Interconversion of the kinetic states of PEP carboxylase is not solely a response to light-to-dark changes but rather a transition that coincides with changes between acidification and deacidification of the cytoplasm. Since malate is released from the vacuole during the deacidification phase, it is suggested that malate sensitization of PEP carboxylase may be caused by malic acid itself (Winter, 1982a). Marked changes in the properties of PEP carboxylase from various CAM species have been shown to occur after isolation (Winter, 1982b); for example, in the enzyme extracted from *Bryophyllum fedtschenkoi* the malate concentration required to inhibit activity by 50% at pH 7.0 was increased by approximately 50-fold after purification (Jones et al., 1981). Such changes suggest that PEP carboxylase from CAM species can exist in more than one stable form; however, interconversion between these kinetic states remains to be convincingly demonstrated *in vitro*.

8.5. PHASE 3—MALIC ACID DECARBOXYLATION IN THE LIGHT

In Phase 3 the stomata remain closed and carbon assimilation via the C3 pathway is dependent on CO_2 supplied by the decarboxylation of malic acid. The inhibition of PEP carboxylase during deacidification prevents competition and futile cycling between the C3 and C4 pathways for the malate derived carbon. The movement of malic acid from the vacuole is passive and high rates of decarboxylation are necessary to maintain the gradient in malate concentration between the vacuole and cytoplasm to facilitate effective release of malate. Since an excessive buildup of CO_2 is inhibitory to the decarboxylating enzymes, the rate of CO_2 fixation and the rate of CO_2 released are coordinated. The rates of photosynthesis and malic acid decarboxylation are interdependent such that the addition of DCMU, which blocks electron transport, also prevents malic acid consumption during phase 3 (Nishida and Hayashi, 1980). Similarly, light intensity has a distinct effect on the rate and extent of malic acid decarboxylation.

The depletion of malic acid is closely correlated with starch formation such that ^{14}C incorporated into malic acid in the dark is quantitatively converted to sugars and glucan in the light. [The reciprocal synthesis and degradation of malate and starch has been used as a definition of CAM (Kluge and Ting, 1978).] Conversion of malic acid to carbohydrate involves decarboxylation and refixation of CO_2 in a closed system because the high intercellular CO_2 concentrations cause tight closure of the stomata. This also leads to an increase in the intercellular O_2 concentration (because of O_2 evolution) but as in C4 plants the high CO_2 level permits CO_2 fixation via RuBP carboxylase with little competing oxygenase activity. C4 and CAM plant have similar systems for decarboxylation (see Section 7.8). A substantial proportion of the C3 residues produced as a result of the decarboxylation reactions can be converted to PEP and used in starch synthesis

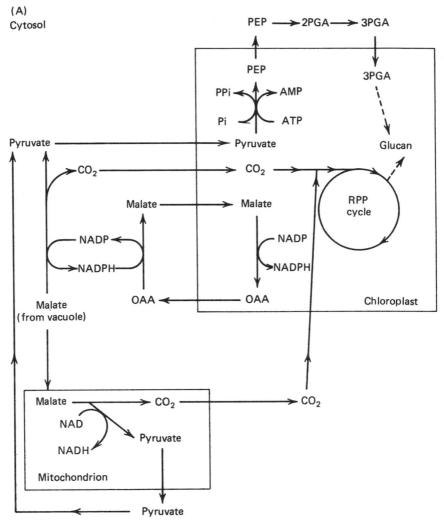

Figure 8.4.   Schemes of malate decarboxylation and glucan synthesis in (A) malic enzyme containing CAM plants.

via gluconeogenesis. The C4 acid decarboxylation enzymes known to be important in CAM are PEP carboxykinase [reaction (7.9)], NADP-malic enzyme [reaction (7.6)], and NAD-malic enzyme [reaction (7.7)]. CAM plants can be divided into two groups by the type of enzyme they contain. Members of the malic enzyme group contain NADP- and/or NAD-malic enzyme but no PEP carboxykinase. In this group the product of decarboxylation is pyruvate and this is converted to PEP (Figure 8.4A) by the action

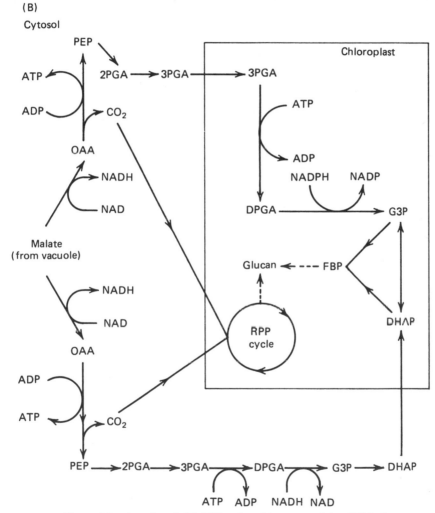

Figure 8.4. (*continued*) (B) PEP carboxykinase containing CAM plants.

of pyruvate Pi dikinase [reaction (7.2)]. The PEP carboxykinase group contains high activities of this enzyme that are far in excess of the activities of the malic enzymes. In PEP carboxykinase containing species PEP is produced directly by the action of this enzyme (Figure 8.4B) and these plants do not need nor contain pyruvate Pi dikinase since PEP carboxykinase effectively bypasses the endothermic conversion of pyruvate to PEP. During the conversion of malic acid to carbohydrate approximately 75% must be derived from PEP directly. PEP produced as a result of the decarboxylation processes is converted to glucan via gluconeogenesis for use in the following dark period, instead of being converted to other products such as sucrose. The remaining 25% of the glucan stored in darkness is provided by refixation of the CO_2 in the light by the C3 pathway.

8.6. PHASE 4—THE LATE LIGHT PERIOD

Phase 4 commences toward the end of the light period, after noon, when the stomata may be opened in plants with access to a sufficiently plentiful supply of water. Following the consumption of the nocturnal store of malic acid, the stomata open and external CO_2 may be assimilated via the C3 pathway. The contribution of C3 photosynthesis in phase 4 to the carbon assimilation of CAM plants varies greatly between species and is dependent on the intensity of irradiance and duration of the light period, which control the rate of malate consumption. When the malate store is depleted and CO_2 supplied from malate no longer saturates photosynthesis, external CO_2 can be assimilated. CO_2 assimilation from the air appears to occur largely by the C3 pathway. In this phase the first stable products of $^{14}CO_2$ fixation are largely PGA and phosphorylated compounds although some malate is synthesized. The rate of CO_2 assimilation can be increased in phase 4 by reducing the O_2 concentration of the surrounding air from 21% to 4% just as it is in C3 plants (Osmond and Bjorkman, 1975), confirming the operation of RuBP carboxylase mediated CO_2 fixation (CO_2 fixation via PEP carboxylase is insensitive to changes in O_2 concentration in this range). In phase 4 the compensation point is increased to a value similar to that displayed by C3 species (50 ppm), which is clearly distinct from the compensation point of near zero ppm that is evident during dark CO_2 fixation in CAM plants (Osmond and Bjorkman, 1975). Malic acid is not generally produced in substantial amounts but external CO_2 may be transferred into malate particularly toward the end of phase 4 (Kluge, 1969; Osmond and Allaway, 1974).

As in phase 2, the path of carbon flow in phase 4 is not constant and there appears again to be a shift in the relative activities of the C3 and C4 pathways. During the initial stages of phase 4 the C3 pathway dominates carbon flow but it gradually changes so that at the end of phase 4 the C4 pathway predominates. During this time the malic acid content is low and this may allow a change in the kinetic properties of PEP carboxylase such that effective malic acid synthesis is again possible. At the end of phase 4 the path of carbon flow is similar to that occurring in phase 1 but because of illumination photosynthetic carbon fixation can still proceed by the C3 pathway and the products may contribute to PEP synthesis. The malic acid formed in phase 4 can be labeled both in the C1 and C4 carboxyl carbons (Osmond and Allaway, 1974) indicating that the double carboxylation can occur in the light in CAM plants, but not in the dark. Phosphoglycerate kinase functions predominantly in the gluconeogenic direction in phase 4 during which malic acid produced from PEP may be derived from PGA possibly as follows:

$$\text{PGA} \xrightarrow{\text{phosphogly-}\atop\text{ceromutase}} \text{2-phosphoglycerate} \qquad (8.2)$$

$$\text{2-phosphoglycerate} \xrightarrow{\text{enolase}} \text{PEP} \qquad (8.3)$$

$^{14}CO_2$ fixation in the light is stimulated by PEP and by 2-phosphoglycerate but is not stimulated by PGA.

Relatively little of the $^{14}CO_2$ fixed in phase 4 is converted into glucan; most of the labeled carbohydrate is found to be sucrose. The switch from glucan synthesis to sucrose synthesis occurs after deacidification. In the absence of the gluconeogenic flux of carbon flow from PEP or pyruvate the net movement of PGA from the cytoplasm to the chloroplast is probably superceded by a net export of triose phosphate from the chloroplast (in exchange for Pi), which is then available for sucrose synthesis. A low stromal [PGA]:[Pi] ratio would tend to inhibit ADP–glucose pyrophosphorylase activity and hence prohibit starch synthesis.

8.7. METABOLITE TRANSPORT

In the CAM cycle there is a complex transport of metabolites between the chloroplasts, mitochondria, and the vacuole both in the conversion of starch and CO_2 to malate in the nocturnal phase and also in the metabolism of malate to starch. Malic enzyme containing species probably possess similar exchange systems for pyruvate, Pi, and PEP as the C4 mesophyll chloroplasts. The presence of a 32,000 molecular weight polypeptide in the chloroplast envelopes of *M. crystallinium* in the CAM mode but not in the C3 mode may be related to the specific transport processes (such as pyruvate and PEP) required in CAM (Edwards et al., 1982). NADP-malic enzyme is localized in the cytoplasm, NAD-malic enzyme in the mitochondria and pyruvate Pi dikinase in the chloroplasts with phosphoglyceromutase and enolase in the cytoplasm. In PEP carboxykinase containing species this enzyme is localized in the cytoplasm. A complex stoichiometry of the phosphate translocator and mitochondrial oxidation of cytoplasmic NADH may be necessary to sustain decarboxylation and gluconeogenesis (Osmond and Holtum, 1982).

The carbohydrate source for nocturnal PEP synthesis has been ascribed to starch glucans and soluble sugars. Starch is localized in the chloroplast but the compartmentation of the glucans and soluble sugars is not known. The chloroplast envelope of *Sedum* species is apparently able to translocate hexose monophosphates. When hexose monophosphates are added to isolated *Sedum* chloroplasts concentrations as low as 50 μM have been found to stimulate CO_2 assimilation (Piazza et al., 1982). The vacuole is the storage site for the large transient pool of malate in CAM cells acting as a sink for malate during nocturnal CO_2 fixation and also as a source of carbon in the light. In addition to malate the vacuoles from *Sedum* appear to contain a transient pool of Pi (Black et al., 1982). The Pi content of the vacuoles was found to be lower at night than late in the day and moved with time in a reciprocal fashion to the diurnal movement of malate. A substantial amount of Pi was found to move out of the vacuole at night and it was estimated that this could cause an increase in the level of Pi in the cytoplasm of approximately 1–2 mM (Black et al., 1982). This is sufficient to have a considerable effect on transport and other metabolic processes and may explain why *Sedum* chloroplasts do not show complete inhibition of photosynthesis in the presence of exogenous Pi concentrations as high as 25 mM. Spalding and Edwards (1980) obtained evidence for the presence of a functioning Pi translocator in the chloroplast envelope of *S. praealtum*. Isolated chloroplasts from CAM plants show very short induction periods even at high concentrations of Pi and also high rates of photosynthesis in the absence of added Pi. The ability of CAM chloroplasts to photosynthesize in the absence of external Pi could reflect the presence of high stromal Pi reserve. Alternatively, since there is limited export of triose phosphate from the chloroplasts in the absence of external Pi, it may reflect the capacity of these chloroplasts to synthesize starch at much higher rates than do C3 chloroplasts. The import/export patterns in CAM chloroplasts in the dark appear to be very similar to the phosphate and dicarboxylate translocator-linked shuttle systems that operate in C3 chloroplasts.

REFERENCES

Black, C. C., Carnal, N. W., and Kenyon, W. H. (1982). In *Crassulacean Acid Metabolism* (I. P. Ting and M. Gibbs, eds.), pp. 51–68. Waverly Press, Baltimore.

Buser-Suter, C., Wiemken, A., and Matile, P. (1982a). *Plant Physiol.* **69**, 456–459.

Buser-Suter, C., Wiemken, A., and Matile, P. (1982b). *Plant Physiol.* **69**, 462–466.

Carnal, N. W. and Black, C. C. (1983). *Plant Physiol.* **71**, 150–155.

Cseke, C., Weeden, N. F., Buchanan, B. B., and Uyeda, K. (1982). *Proc. Nat. Acad. Sci. USA* **79**, 4322–4362.

Cockburn, W., Ting, I. P., and Stemberg, L. O. (1979). *Plant Physiol.* **63**, 1029–1032.

Edwards, G. E., Foster, J. G., and Winter, K. (1982). In *Crassulacean Acid Metabolism* (I. P. Ting and M. Gibbs, eds.), pp. 92–111. Waverly Press, Baltimore.

Jones, R., Buchanan, I. C., Wilkins, M. B., Fewson, C. A., and Malcolm, A. D. B. (1981). *J. Exp. Bot.* **32**, 427–441.

Kelly, G. J. and Latzko, E. (1976). *FEBS Lett.* **68**, 55–58.

Kluge, M. (1969). *Planta* **88**, 113–129.

Kluge, M., Fischer, A., and Buchanan-Bollig, I. C. (1982). In *Crassulacean Acid Metabolism* (I. P. Ting, and M. Gibbs, eds.), pp. 31–50. Waverly Press, Baltimore.

Kluge, M. and Ting, I. P. (1978). In *Crassulacean Acid Metabolism*. An analysis of an ecological adaptation. Springer-Verlag, New York.

Luttge, U. and Ball, E. (1978). *Z. Pflanzenphysiol.* **90**, 69–77.

Nishida, K. and Hayashi, Y. (1980). *Plant Sci. Lett.* **19**, 271–276.

Nobel, P. S. and Hartsock, T. L. (1983). *Plant Physiol.* **71**, 71–75.

Osmond, C. B. (1978). *Ann. Rev. Plant Physiol.* **29**, 379–414.

Osmond, C. B. (1982). IN *Crassulacean Acid Metabolism* (I. P. Ting, and M. Gibbs, eds.), pp. 112–127. Waverly Press, Baltimore.

Osmond, C. B. and Allaway, W. G. (1974). *Aust. J. Plant Physiol.* **1**, 503–511.

Osmond, C. B. and Bjorkman, O. (1975). *Aust. J. Plant Physiol.* **2**, 155–162.

Osmond, C. B. and Holtum, J. A. M. (1982). In *The Biochemistry of Plants*, Vol. 8 (M. D. Hatch and N. K. Boardman, eds.), pp. 283–328. Academic Press, New York.

Piazza, G. J., Smith, J. G., and Gibbs, M. (1982). *Plant Physiol.* **70**, 1748–1758.

Powles, S. B., Chapman, K. S. R., and Osmond, C. B. (1980). *Aust. J. Plant Physiol.* **7**, 737–747.

Queiroz, O. and Brulfert, J. (1982). In *Crassulacean Acid Metabolism* (I. P. Ting and M. Gibbs, eds.), pp. 203–230. Waverly Press, Baltimore.

Spalding, M. H. and Edwards, G. E. (1980). *Plant Physiol.* **65**, 1044–1048.

Turner, J. P. and Turner, D. H. (1980). In *The Biochemistry of Plants* Vol. 2, pp. 279–316. Academic Press, New York.

Von Willert, D. J., Brinckman, E., Scheitler, B., Thomas, D. A., and Treichel, S. (1979). *Planta* **147**, 31–36.

Warren, D. M. and Wilkins, M. B. (1961). *Nature* **191**, 686–688.

Wilkins, M. B. (1969). In *Physiology of Plant Growth and Development* (M. B. Wilkins, ed.), pp. 647–671. McGraw-Hill, London.

Winter, K. (1982a). In *Crassulacean Acid Metabolism* (I. P. Ting and M. Gibbs, eds.), pp. 153–169. Waverly Press, Baltimore.

Winter, K. (1982b). *Planta* **154**, 298–309.

Winter, K. and Tenhunen, J. D. (1982). *Plant Physiol.* **70**, 1718–1722.

Winter, K. and Von Willert, D. J. (1972). *Z. Pfanzenphysiol.* **67**, 166–170.

9

PHOTORESPIRATION

9.1. INTRODUCTION

In darkness net O_2 uptake and CO_2 release, resulting from mitochondrial respiration, is observed in photosynthetic cells. The pattern of gas exchange is different in CAM chlorenchyma cells because CO_2 evolved by mitochondrial respiration is masked by net CO_2 uptake that occurs in the nocturnal carboxylation phase. Upon illumination net O_2 uptake in C3, C4, and CAM plants is rapidly superceded by oxygen evolution arising from water oxidation by the photosynthetic electron transport chain. The rate of oxygen evolution during photosynthesis is generally 100-fold greater than the rate of dark O_2 uptake. However, in the light, plants exhibit O_2 uptake and CO_2 release associated with the synthesis and metabolism of glycollate (Tolbert, 1981; Lorimer and Andrews, 1973). This second type of plant respiration is light dependent and is intimately associated with photosynthesis. It is, therefore, known as photorespiration. The term *photorespiration* is sometimes used to include all types of O_2 uptake in the light, for example, O_2 taken up in pseudocyclic electron flow but strictly should refer only to gas exchange associated with glycollate metabolism. The majority of O_2 uptake in the light results from the oxygenation of RuBP via the oxygenase reaction of RuBP carboxylase [reaction (5.14)], which produces PGA and 2-phosphoglycollate. In the photorespiratory pathway of phosphoglycollate reassimilation CO_2 is lost and there is no energy conservation as net ATP or NADPH, only energy consumption. The photorespiratory cycle serves to recover carbon by converting phosphoglycollate ultimately back to PGA (Figure 9.1). Approximately 75% of the carbon in phosphoglycollate is recovered in this way, while the remaining 25% of glycollate carbon is lost as CO_2 in the mitochondria. In contrast to mitochondrial respiration, photorespiration does not conserve energy but consumes it. Photorespiration is an irreversible exothermic process and has a high demand for the energy captured by photosynthesis. In C3 photosynthesis the energy required to drive CO_2 fixation is dependent on the CO_2 and O_2 concentrations present in the intercellular spaces of the leaves. In the absence of photorespiration, 3 mole ATP and 2 mole NADPH are required per CO_2 fixed and the quantum requirement is significantly less than that of CO_2 fixation by the C4 pathway. However, during illumination in normal air, which contains 21% O_2 and 300 ppm (0.33%) CO_2, the mesophyll cells of C3 plants loose a significant proportion of their newly assimilated carbon (30–60%) through the photorespiratory cycle and have a lower quantum yield than that measured in

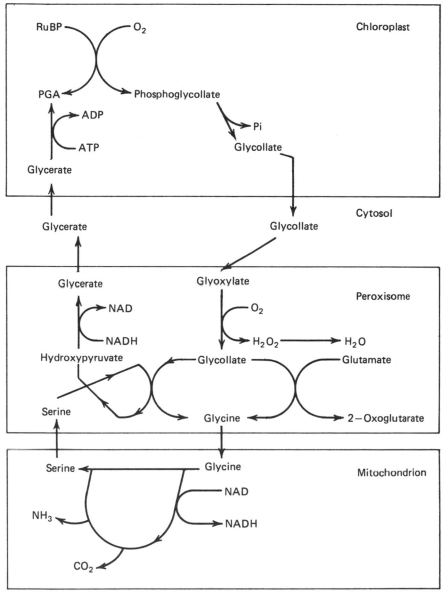

Figure 9.1. The compartmentation of the reactions of the photorespiratory cycle in the chloroplasts, cytosol, peroxisomes, and mitochondria.

199

the presence of low O_2. At low CO_2 concentrations ($\sim$100 ppm) and 21% O_2 a major proportion of carbon flow is diverted through the photorespiratory cycle increasing the quantum requirement by approximately 70%. The rate of energy consumption is therefore dependent on the relative O_2 and CO_2 concentrations. However the stoichiometry of ATP and NADPH requirements is not significantly altered by the diversion of carbon through the photorespiratory cycle. In C4 and CAM species photorespiration is suppressed because of the CO_2 concentrating mechanisms and any CO_2 lost by the photorespiratory cycle is effectively recaptured by the action of PEP carboxylase. C4 and CAM plants therefore show no apparent photorespiration but they do possess the complete photorespiratory cycle, although the activities of some of the component enzymes are diminished compared to C3 species. The compensation point in these plants is near zero. The CO_2 concentrating mechanisms in C4 and CAM plants that serve to prohibit oxygenation of RuBP and consequently to reduce photorespiration are energy requiring and reduce the quantum yield of photosynthesis (Ehleringer and Bjorkman, 1977; Nobel, 1977).

At the compensation point, which is approximately 50 ppm in C3 plants at 25°C and in 21% O_2, the CO_2 lost as a result of photorespiration virtually balances CO_2 fixation by the RPP pathway. At this compensation point the rate of CO_2 fixation proceeds at only half the rate of the oxygenation of RuBP and carbon turnover consumes 65–70% as much energy as net CO_2 fixation in the absence of photorespiration (Osmond, 1981). Photorespiration can be suppressed by a reduction in the partial pressure of oxygen in the surrounding air or by increasing $[CO_2]$. Thus, if the O_2 is decreased to 2% almost no photorespiration is observed and the CO_2 compensation point of C3 leaves can be reduced to almost zero.

The photorespiratory pathway involves reactions that occur in the chloroplasts, peroxisomes, mitochondria, and cytoplasm (Figure 9.1), and close cooperation and regulation of metabolite flow and exchange between these organelles is required. In electron micrographs of the photosynthetic cells of higher plants these organelles are seen to be closely appressed to each other, which may facilitate direct exchange of metabolites and achieve coordinated function. Phosphoglycollate is produced in the chloroplasts and there converted to glycollate, which then leaves the chloroplast. The other cell organelles are involved in the recycling and recovery of the carbon lost as glycollate.

9.2. GLYCOLLATE BIOSYNTHESIS

Phosphoglycollate is the first stable committed substrate of the photores-piratory cycle. It is produced primarily from C1 and C2 of RuBP by the oxygenase reaction of RuBP carboxylase [reaction (5.14)]. The oxygenase reaction was shown by Bowes et al. (1971) to result in the oxygen dependent cleavage of RuBP to yield PGA and phosphoglycollate. When $^{18}O_2$ was supplied to isolated RuBP carboxylase, one atom of $^{18}O_2$ was found to be incorporated into the carboxyl group of phosphoglycollate during oxygenation of RuBP (Lorimer et al., 1973). The oxygenase reaction can account for most if not all of the glycollate found in leaves. At the concentrations of CO_2 and O_2 normally present in the atmosphere oxygenation of RuBP and therefore O_2 inhibition of CO_2 fixation are inevitable because oxygen is a competitive inhibitor with respect to CO_2 (Lorimer and Andrews, 1973). The concentration of dissolved O_2 in water in equilibrium with air is ap-proximately 267 μM at 25°C, while that of CO_2 is approximately 11 μM. The $K_m O_2$ of the oxygenase activity is between 200 and 500 μM, while the $K_m CO_2$ of the carboxylase is generally between 10 and 25 μM. The rate of glycollate synthesis is dependent on the relative concentrations of CO_2 and O_2 in the stroma. Since O_2 is a product of photosynthesis, the chloroplast environment is rich in oxygen but CO_2 is not always readily available in the microenvironment of the intercellular spaces. Because of the lower affinity of the enzyme for O_2 and the lower V_{max} of the oxygenase reaction carboxylation should proceed at least 4 times the rate of oxygenation in C3 plants. The ratio of carboxylase:oxygenase activity of the isolated enzyme assayed with air levels of O_2 and saturating CO_2 favors carboxylation by about 8:1. CO_2 is more soluble in water than O_2 and solubility of both decrease with increasing temperature. However, the solubility of CO_2 de-creases more rapidly than O_2 as temperature increases. Thus, at higher temperatures the ratio of oxygenase to carboxylase activities is increased and the C3 plant is forced into a situation of increased photorespiration relative to CO_2 fixation.

When manganese is used in the activation process of isolated enzyme (see Section 4.4) in place of magnesium, the relative rates of the oxygenase and carboxylase activities are altered often in favor of the former (Wildner and Henkel, 1979; Christeller and Laing, 1979). It has been suggested that the enzyme binds RuBP by the interaction of ionic forces of the phosphate groups with basic amino groups on arginine or lysine residues of the protein. A carbamate group is formed by the activating CO_2 (Lorimer, 1983) exclusively with the e-amino group of a lysyl residue on the large subunits of the enzyme and this in turn binds either magnesium or manganese (Lorimer and Miziorko, 1980). The bound RuBP is then suggested to undergo keto-enoltautomerization so that the C2 carbon becomes anionic and the reaction

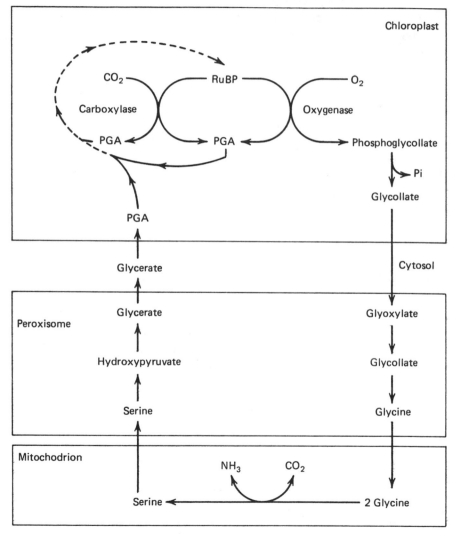

Figure 9.2.   Simplified diagram of the path of carbon flow in photorespiration.

can then proceed with the addition of either O_2 or CO_2 (Figure 9.2). The reason for the different response of the carboxylase and oxygenase reactions to Mn^{2+} is not fully understood but it does show that differential modulation of the two activities may be possible. Unlike other monooxygenase reactions RuBP oxygenase does not show an absolute requirement for a transition metal ion or other redox active prosthetic group. Estimates of the carboxylase:oxygenase ratios in a range of plant and bacterial sources shows that there is a wide variation between species and may suggest that changes in the enzyme structure can effect the relative rates of the alternative reactions. In order to decrease photorespiration it is necessary to suppress the oxygenase activity of RuBP carboxylase independently of the carboxylase, however, specific inhibition of the oxygenase activity has not, as yet, been demonstrated (Keys et al., 1982).

In the chloroplast the phosphate moiety is hydrolyzed from phosphoglycollate by a specific phosphatase (Christeller and Tolbert, 1978). It is likely that the size of the phosphoglycollate pool in the stroma is kept small because this metabolite is a powerful inhibitor of the Calvin cycle enzyme triose phosphate isomerase ($K_i = 0.15 \ \mu M$). It is also a powerful positive effector of the 2,3-diphosphatase activity of 2,3-phosphoglycerate mutase. Phosphoglycollate phosphatase requires divalent cations for activity and is inhibited by RuBP. Some unicellular algae excrete glycollate into the surrounding environment. This represents a direct loss of photosynthate. A small proportion of the total glycollate is synthesized by an alternative mechanism in which $^{18}O_2$ is not incorporated during synthesis. Glycollate may arise from a direct condensation reaction or from an intermediate in the transketolase reaction [reactions (4.8) and (4.11)]. In the latter reaction oxidation of the thiaminpyrophosphate-C_2 complex of transketolase could produce glycollate. In the Calvin cycle transketolase catalyzes the transfer of C2 units between sugar phosphates. The C2 units are associated with a thiaminpyrophosphate cofactor at the active site of the enzyme and may be oxidized nonenzymically by agents such as H_2O_2 to glycollate and thiaminepyrophosphate (Zelitch, 1975).

9.3. THE METABOLISM OF GLYCOLLATE

Glycollate appears to diffuse readily across the chloroplast envelope (Takabe and Akazawa, 1981) and move into the peroxisomes; however, carrier-mediated transport of glycollate has been reported in pea chloroplasts (Howitz and McCarty, 1983). The glycollate level in leaves is generally found to be low since glycollate is rapidly metabolized to glyoxylate in the peroxisomes via the action of glycollate oxidase [reaction (9.1)] a flavine mononucleotide (FMN) containing enzyme:

$$\text{glycollate} \ + \ O_2 \ \longrightarrow \ \text{glyoxylate} \ + \ H_2O_2 \qquad (9.1)$$

The production of glyoxylate is the primary function of this enzyme but it will also catalyze the oxidation of L-lactic acid to pyruvate and glyoxylate to oxalate:

$$\text{glyoxylate} \ + \ O_2 \ \longrightarrow \ \text{oxalate} \ + \ H_2O_2 \qquad (9.2)$$

Glycollate oxidase is cyanide insensitive but may be inhibited by compounds such as 2-hydroxy-3-butynoic acid, which reacts with the FMN prosthetic group and hydroxysulfonates especially pyridyl hydroxymethane sulfonate. Glycollate oxidase is utilized during glycollate metabolism in higher plants and multicellular algae; however, many unicellular algae employ an alternative enzyme, glycollate dehydrogenase, in place of glycollate oxidase. Unlike glycollate oxidase, which utilizes O_2 and produces H_2O_2, glycollate dehydrogenase is not linked to O_2 uptake and is strongly inhibited by cyanide. In glycollate dehydrogenase containing algae the catalase complement is only approximately one-tenth of that found in glycollate oxidase utilizing organisms. Glycollate dehydrogenase is localized both in the peroxisomes and mitochondria of the unicellular algae. Mitochondrial glycollate oxidation appears to be linked to the electron transport chain through a b-type cytochrome and is coupled to ATP synthesis. Glycollate may be excreted from glycollate dehydrogenase containing algal species at times when glycollate synthesis exceeds the capacity of the enzyme activity. Glycollate dehydrogenase oxidises glycollate and also D-lactic acid.

Hydrogen peroxide produced by the glycollate oxidase reaction [reaction (9.1)] may be removed by the action of catalase [reaction (3.6)] or it may react nonenzymically with glyoxylate to release CO_2 from the carboxyl group and produce formate [reaction (9.3)]. In the peroxisome glyoxylate may compete with catalase for the available H_2O_2 but the nonenzymic peroxidation of glyoxylate appears to account for only a small percentage of the CO_2 released in photorespiration:

$$\text{glyoxylate} \ + \ H_2O_2 \ \longrightarrow \ CO_2 \ + \ \text{formate} \ + \ H_2O \qquad (9.3)$$

This reaction occurs to a limited extent in peroxisomes and the formate thus produced may be activated to a tetrahydrofolate derivative by the action of the enzyme formyltetrahydrofolate synthetase and thus enter the folate pool.

9.4. THE SYNTHESIS OF GLYCINE AND SERINE

The conversion of glyoxylate to glycine and serine occurs via the action of glutamate/glyoxylate and serine/glyoxylate aminotransferases, which catalyze essentially irreversible reactions (Figure 9.3). These two aminotransferases are highly specific for glyoxylate, and for their respective amino acid donors although the serine/glyoxylate aminotransferase will also use alanine as an amino donor. The serine/glyoxylate aminotransferase that catalyzes reaction (9.4) is the more active of the two enzymes, in leaf peroxisomes, facilitating the path of carbon to glycine:

$$\text{L-serine} + \text{glyoxylate} \longrightarrow \text{glycine} + \text{hydroxypyruvate}$$
$$(9.4)$$

The formation of serine requires the participation of two glycine molecules, thus necessitating the presence of the second aminotransferase, which uses a glutamate donor:

$$\text{L-glutamate} + \text{glyoxylate} \longrightarrow \text{glycine} + \text{2-oxoglutarate}$$
$$(9.5)$$

Glycine is not further metabolized in the peroxisome but is transferred to the mitochondria where two glycine molecules are converted to serine with the release of CO_2. The oxidative decarboxylation of glycine accounts for a significant proportion of the CO_2 lost during photorespiration. As a neutral amino acid glycine permeates into the mitochondria and should not require active uptake. However, pea leaf mitochondria have been shown to possess a glycine transporter that appears to be exposed on the outer surface of the inner mitochondrial membrane in which it is situated. Glycine transport by this means appears to be driven by the transmembrane proton gradient (Walker et al., 1982). In the mitochondria glycine decarboxylation is facilitated by a membrane bound glycine decarboxylase linked to a serine hydroxymethyl transferase in the matrix (Bird et al., 1972; Moore et al., 1977). The glycine decarboxylase multienzyme complex oxidatively decarboxylates glycine

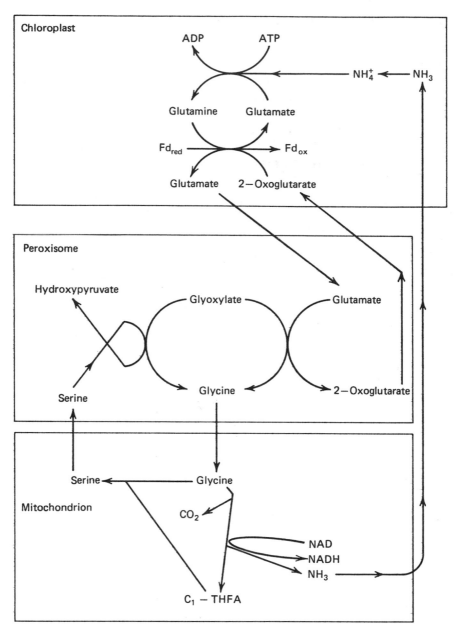

Figure 9.3. The conversion of glycine to serine showing the conservation of aminogroups.

releasing CO_2, from the carbonyl carbon, and ammonia and produces a 5,10-methylene tetrahydrofolic acid (C1-THFA) complex:

glycine + tetrahydrofolate + NAD^+ $\rightleftharpoons$
$$5,10\text{-methylenetrahydrofolate} + CO_2 + NH_3 + NADH \quad (9.6)$$

This reaction is associated with electron transport chain and ATP synthesis in the mitochondria such that two molecules of glycine are decarboxylated per O_2 taken up. This linkage to electron transport limits the reversibility of the reaction. The C1-THFA generated in the mitochondria from glycine or serine is the precursor for all the essential C1 requiring reactions in plants. Apart from serine synthesis C1-THFA can be used for all synthesis utilizing C1 groups, for example, the production of purines and methionine. Compounds that may be used to inhibit the conversion of glycine to serine include the competitive inhibitors, aminoacetonitrile (K_i = 126 μM) and glycine hydroxamate (K_i = 240 μM), and also isonicotinic acid hydrazide. In the photorespiratory cycle the major proportion of the methylenetetrahydrofolic acid is condensed with a second glycine molecule to form serine according to reaction (9.7) catalyzed by the enzyme serine hydroxymethyltransferase:

glycine + 5,10-methylenetetrahydrofolate $\rightleftharpoons$
$$\text{L-serine} + \text{tetrahydrofolate} \quad (9.7)$$

Stoichiometric amounts of NH_3 and CO_2 are released via the glycine decarboxylase system. Ammonia is highly toxic and must be quickly refixed into glutamate or glutamine. The amino group conserved in glutamate may then be shuttled back to the peroxisome for the aminotransferase reaction (Figure 9.3). The NAD requirement of the glycine decarboxylase reaction is probably supplied by the presence of a decarboxylate shuttle across the inner mitochondrial membrane coupled to reduction of oxaloacetate to malate by NAD-malate dehydrogenase in the matrix.

9.5. THE GLYCERATE PATHWAY

Serine produced in the mitochondria can be subsequently returned to the peroxisomes to enter the reversible glycerate pathway by which phosphoglycerate is regenerated and returned to the RPP pathway. Hydroxypyruvate produced from serine via the action of serine/glyoxylate aminotransferase [reaction (9.4)] can be reduced to glycerate via the action of glycerate dehydrogenase:

$$\text{hydroxypyruvate} + \text{NADH} + \text{H}^+ \rightleftharpoons \text{D-glycerate} + \text{NAD}^+$$

$$(9.8)$$

This enzyme will also catalyze the reduction of glyoxylate to glycollate but the affinity for glyoxylate is much less ($K_m = 20$ mM) than the affinity for hydroxypyruvate ($K_m = 0.2$ mM). The NADH required for this reaction may possibly be provided by the action of malate dehydrogenase. The glycerate produced from hydroxypyruvate via reaction (9.8) can then be phosphorylated by glycerate kinase most of which is localized in the chloroplast. The inner envelope membrane of the chloroplast prevents the uncontrolled movement of weak acids into the chloroplast, since this would lead to an acidification of the stroma that is inhibitory to CO_2 fixation (Enser and Heber, 1980). The inner envelope membrane of spinach chloroplasts has been shown to contain a specific glycerate translocator (Robinson, 1982a). The activity of this carrier is stimulated by light and it is possible that the glycerate anion enters the chloroplast in symport with protons (Robinson, 1982b). As a result of the action of the glycerate translocator glycerate accumulation may be achieved in the stroma. This favors the activity of the glycerate kinase, which generates PGA from glycerate and does not have a high affinity for its substrate (Heber et al., 1974). In this manner phosphoglycerate may be recycled and enter the RPP pathway.

9.6.   REASONS FOR PHOTORESPIRATION

The process of photorespiration is apparently ubiquitous among plants yet the function of this energetically expensive pathway is not immediately apparent. In addition, carbon flow through the photorespiratory cycle can be decreased by a reduction in oxygen concentration or an increase in the CO_2 concentration of the air surrounding the plant without immediate adverse effects on metabolism. In these circumstances suppression of photorespiration is actually beneficial in terms of increased yields. It can be argued that the photorespiratory cycle may provide a compensatory mechanism for a major deficiency in the enzyme RuBP carboxylase, which may be considered to be more suited for operation in the primeval CO_2 rich reducing environment in which it presumably initially evolved. For mechanistic reasons oxygenation of RuBP is an unavoidable side reaction to carboxylation (Andrews and Lorimer, 1978) and photorespiration may then be seen to serve as a recycling mechanism for carbon lost from the Calvin cycle, compensating for the shortcomings of RuBP carboxylase. However, the energy consumption in photorespiration may be used to the advantage of the plant protecting against photooxidative damage. Light energy that cannot be used effectively in photosynthesis may give rise to destructive side reactions which would be deleterious to plant metabolism. Photorespiration may provide a protective mechanism utilizing energy and recycling CO_2 in conditions where the concentration of CO_2 is low relative to O_2, thus maintaining CO_2 in the cellular environment. Conditions of CO_2 depletion often occur in leaves in the light. Similarly, in full sunlight a water deficit may be encountered because water uptake by the roots fails to keep pace with the rate of transpiration. This results in stomatal closure and the CO_2 level in the intercellular spaces may then decrease to the compensation point. The photorespiratory process may then allow the electron transport chain to function, since the NADPH and ATP pools are turned over, while the net gas exchange is severely limited.

Photoinhibition has been demonstrated in isolated thylakoids and chloroplasts, algae, and leaves (Osmond, 1981; Powles and Osmond, 1978). Maximal photoinhibition is induced when leaves are illuminated in the absence of CO_2 and at low (1%) O_2 such that carbon metabolism and electron transport to O_2 are severely restricted. Photoinhibition in leaves can be wholly or partially prevented by increasing the O_2 concentration to 21% thus allowing photorespiration to proceed (Powles and Osmond, 1978). Under limiting CO_2 and 21% O_2, isolated intact chloroplasts will convert almost all of their endogenous Calvin cycle intermediates to glycollate, which is then excreted from the chloroplasts (Kirk and Heber, 1976). Under conditions that favor glycollate synthesis the Pi optimum for photosynthetic O_2 evolution in isolated wheat chloroplasts was found to be dramatically reduced such that the optimal Pi level was between 0 and 25 μM at 0.3 mM bicarbonate (Usuda and Edwards, 1982). This may be explained by the fact that glycollate synthesis is not a Pi-consuming process as is the synthesis of triose phosphate. Krause et al. (1978) showed that illumination of isolated chloroplasts in the absence of CO_2 produces irreversible inhibition of CO_2 fixation. In contrast to the situation in leaves, inhibition in chloroplasts has been shown to be dependent on the presence of strong light and also the presence of O_2 (Cornic et al., 1982). The extent of photoinhibition is dependent on the concentration of CO_2 present during the light period and could be totally prevented by 40 μM CO_2. In leaves a CO_2 concentration corresponding to the intracellular partial pressure at the compensation point in air is sufficient to prevent photoinhibition. In chloroplasts photoinhibition comprises of two separate components, an inhibition of electron transport and also an inhibition of CO_2 fixation. Direct photoinhibition of CO_2 fixation at 21% O_2 in the absence of CO_2 may be caused by oxidative processes (see Section 3.4) and also to the removal of activating CO_2 from RuBP carboxylase. The O_2 dependent process of photoinhibition is largely independent of the rate of photosynthesis and occurs throughout the strong light treatment (Cornic et al., 1982). Photoinhibition is observed when plants adapted to growth at low light intensities are subject to strong illumination (Powles and Critchley, 1980). In low-light grown plants, the processes of photo-inhibition are similar to those that occur in high light plants, exposed to high irradiation in the absence of CO_2. With low-light grown plants photoinhibition cannot be completely prevented by the presence of CO_2 and/or O_2. Photoinhibition of electron transport in these plants appears to occur primarily at the photochemical reaction centers of PSII such that following photoinhibitory damage electron transport associated with PSII is significantly decreased while PSI electron transport is unchanged. The sensitive site does not appear to be the water-splitting complex but may be located closer to the reaction center on the donor side of PSII (Critchley, 1981). Secondary processes involving O_2 uptake may subsequently cause photooxidative damage as a result of the formation of highly active oxygen species such as singlet O_2 or hydroxyl radicals.

REFERENCES

Andrews, T. J. and Lorimer, G. H. (1978). *FEBS Lett.* **90**, 1–9.

Bird, I. F., Corneliu, M. J., Keys, A. J., and Whittingham, C. P. (1972). *Phytochemistry* **11**, 1587–1594.

Bowes, G., Ogren, W. L., and Hageman, R. H. (1971). *Biochem. Biophys. Res. Commun.* **45**, 716–722.

Christeller, J. T. and Laing, W. (1979). *Biochem. J.* **183**, 747–750.

Christeller, J. T. and Tolbert, N. E. (1978). *J. Biol. Chem.* **253**, 1780–1785.

Cornic, G., Woo, K. C., and Osmond, C. B. (1982). *Plant Physiol.* **70**, 1310–1315.

Critchley, C. (1981). *Plant Physiol.* **67**, 1161–1165.

Ehleringer, J. and Bjorkman, O. (1977). *Plant Physiol.* **59**, 86–90.

Enser, U. and Heber, U. (1980). *Biochim. Biophys. Acta* **592**, 577–591.

Heber, U., Kirk, M. R., Gimmler, H., and Schafer, G. (1974). *Planta* **120**, 31–46.

Howitz, K. T. and McCarty, R. E. (1983). *FEBS Lett.* **154**, 339–342.

Keys, A. J., Bird, I. F., and Cornelius, M. J. (1982). In *Chemical Manipulation of Crop Growth and Development* (J. S. Mclaren, ed.), pp. 39–53. Butterworths, London.

Kirk, M. R. and Heber, U. (1976). *Planta* **132**, 131–143.

Krause, G. H., Kirk, M., Heber, U., and Osmond, C. B. (1978). *Planta* **142**, 229–233.

Ku, S. B. and Edwards, G. E. (1977). *Plant Physiol.* **59**, 986–990.

Lorimer, G. H. (1983). *Trend. Biochem. Sci.* **8**, 65–68.

Lorimer, G. H. and Andrews, T. J. (1973). *Nature* **243**, 359–360.

Lorimer, G. H., Andrews, T. J., and Tolbert, N. E. (1973). *Biochemistry* **12**, 18–23.

Lorimer, G. H. and Miziorko, H. M. (1980). *Biochemistry* **19**, 5321–5328.

Moore, A. L., Jackson, C., Halliwell, B., Dench, J. E., and Hall, D. O. (1977). *Biochem. Biophys. Res. Commun.* **78**, 483–491.

Nobel, P. S. (1977). *Oecolgia* **27**, 117–133.

Osmond, C. B. (1981). *Biochim. Biophys. Acta* **639**, 77–98.

Powles, S. B. and Critchley, C. (1980). *Plant Physiol.* **65**, 1181–1187.

Powles, S. B. and Osmond, C. B. (1978). *Aust. J. Plant Physiol.* **5**, 619–629.

Robinson, S. P. (1982a). *Biochem. Biophys. Res. Commun.* **106**, 1027–1034.

Robinson, S. P. (1982b). *Plant Physiol.* 70, 1032–1038.

Takabe, T. and Akazawa, T. (1981). *Plant Physiol.* **68**, 1093–1097.

Tolbert, N. E. (1981). In *The Biochemistry of Plants*, Vol. 2, Metabolism and Respiration (D. D. Davies, ed.), pp. 488–525. Academic Press, New York.

Usada, H. and Edwards, G. E. (1982). *Plant Physiol.* **69**, 469–473.

Walker, G. H., Sarojini, G., and Oliver, D. J. (1982). *Biochem. Biophys. Res. Commun.* **107**, 856–861.

Wildner, G. F. and Henkel, J. (1979). *Planta* **146**, 223–228.

Zelitch, I. (1975). *Ann. Rev. Biochem.* **44**, 123–145.

INDEX